AF845059

Swarm Intelligence Applications for the Cities of the Future

Editors
Leonor Marques Mano Domingos
ISCTE - University Institute of Lisbon
Lisboa, Portugal

Maria José Sousa
ISCTE - University Institute of Lisbon
Lisbon, Portugal

CRC Press
Taylor & Francis Group
Boca Raton London New York

CRC Press is an imprint of the
Taylor & Francis Group, an **informa** business
A SCIENCE PUBLISHERS BOOK

This cover has been created using images generated using OpenArt.AI

First edition published 2025
by CRC Press
2385 NW Executive Center Drive, Suite 320, Boca Raton FL 33431

and by CRC Press
4 Park Square, Milton Park, Abingdon, Oxon, OX14 4RN

© 2025 Leonor Marques Mano Domingos and Maria José sousa

CRC Press is an imprint of Taylor & Francis Group, LLC

Reasonable efforts have been made to publish reliable data and information, but the author and publisher cannot assume responsibility for the validity of all materials or the consequences of their use. The authors and publishers have attempted to trace the copyright holders of all material reproduced in this publication and apologize to copyright holders if permission to publish in this form has not been obtained. If any copyright material has not been acknowledged please write and let us know so we may rectify in any future reprint.

Except as permitted under U.S. Copyright Law, no part of this book may be reprinted, reproduced, transmitted, or utilized in any form by any electronic, mechanical, or other means, now known or hereafter invented, including photocopying, microfilming, and recording, or in any information storage or retrieval system, without written permission from the publishers.

For permission to photocopy or use material electronically from this work, access www.copyright.com or contact the Copyright Clearance Center, Inc. (CCC), 222 Rosewood Drive, Danvers, MA 01923, 978-750-8400. For works that are not available on CCC please contact mpkbookspermissions@tandf.co.uk

Trademark notice: Product or corporate names may be trademarks or registered trademarks and are used only for identification and explanation without intent to infringe.

Library of Congress Cataloging-in-Publication Data (applied for)

ISBN: 978-1-032-65675-5 (hbk)
ISBN: 978-1-032-65677-9 (pbk)
ISBN: 978-1-032-65678-6 (ebk)

DOI: 10.1201/9781032656786

Typeset in Times New Roman
by Prime Publishing Services

Acknowledgements

The editors would like to extend their deepest gratitude to the Scientific Committee whose invaluable expertise and unwavering dedication have made this book possible. Their insightful guidance, thorough reviews, and constructive feedback have significantly enriched the quality of this work. The editors would like to specifically acknowledge: Dr. Filipa Osório, specialist in Digital Architecture and Parametric Design; Prof. David Ferraz, expert in Public Policies, Sociology, Economics, and Management; Prof. Helena Belchior, specialist in Environment, Sustainability, Social Policies, and Education; Dr. Pedro Mariano, expert in Optimization Algorithms, Artificial Intelligence, and Machine Learning; and Prof. João Pavia, expert in Big Data, Machine Learning, and Cybersecurity.

A special thank you to Prof. Vasco Rato for his support and perspective, for which the editors are profoundly grateful and, also, for writing the foreword for this book. The editors also extend their appreciation to the contributing authors whose innovative research and commitment to excellence have been fundamental to this publication. Additionally, heartfelt thanks go to the technical staff and administrative team, whose hard work behind the scenes ensured the smooth progression of this project. Their support and dedication have been instrumental in bringing this book to fruition.

Lastly, the editors wish to thank their families and friends for their unwavering support. Their encouragement and patience have been a constant source of motivation. Thank you to everyone who contributed to making this book a reality.

Foreword

Since ancient times, cities have been a center of social, economic, and cultural governance. The very diverse structures that have been developed to manage the common space have depended, to a greater or lesser extent, on the ability to seek and maintain consensus. In the governance models characteristic of contemporary liberal democracies, the notion of representativeness is decisive. Proximity between representatives and those they represent plays a critical role as an instrument for legitimizing the exercise of power, i.e., public policies that are designed and implemented with the aim of improving people's well-being within the context of available resources. In cities, this proximity is particularly important because, by definition, due to the territorial scale involved, this is the level of decision-making with the most immediate impact on people's lives.

Today, cities are facing huge challenges associated with, among other things, climate change, the scarcity of resources, the loss of biodiversity, social phenomena such as migration, the lack of affordable housing, or major changes in the structure of the economy and employment. On the other hand, the technological development of the last decade has brought with it a whole new capacity to produce and distribute content very easily and very quickly. The fact that, in general, this content is not curated, makes it possible to construct, sometimes legitimately and sometimes illegitimately, different narratives of the same reality. In this context, those governance models are being questioned. Citizens increasingly seem to perceive that they are not properly represented by elected representatives, which leads to a progressive distancing from those who govern. Thus, a model that aims to nurture consensus for a balanced management of the common good, with respect for all people, tends to lose meaning.

The application of co-governance models will certainly be part of the solution. Various experiments have been carried out, generally successfully. However, if the governance model is gradually changed, the ability to negotiate and generate consensus must always be ensured. This is, after all, about strong coexistence between people who share a territory and who have to find ways of managing it together.

Recent developments allow, today, for a fantastic capacity to collect and analyze information, giving it customized meaning. So the question seems obvious: what is the role of the latest technology, and its future developments in creating new opportunities for mediation that contribute to once again ensuring common ground for debate and consensus, in the legitimate aspiration for better living conditions, within the limits of ecosystems to support life on Earth and, in particular, in cities?

This book aims to contribute to answer this question.

Swarm intelligence explores and draws potential from collective and decentralized behavior. It has a scalable nature and the ability to deal with complex and dynamic systems such as cities. Its application will be part of the future of urban management, directly involving citizens and contributing to more resilient and inclusive cities.

Vasco Rato
May, 30th 2024

Preface

As the world accelerates into an era of rapid urbanization and technological innovation, the cities of tomorrow are being shaped by forces that challenge traditional paradigms of urban planning, governance, and social interaction. This book, *Swarm Intelligence Applications for the Cities of the Future*, emerges from the intersection of architecture, public policies, and digital technology—fields that are increasingly intertwined as we seek to build cities that are not only smarter but also more sustainable, resilient, and equitable.

As editors, we have curated a collection of cutting-edge insights that address some of the most pressing challenges facing future cities. At its core is the concept of swarm intelligence, rooted in the collective behaviour of decentralized systems. This book explores how these principles can be harnessed to create urban environments that are more adaptive, efficient, and responsive to the needs of their inhabitants.

Chapters cover a diverse array of themes, crucial for the future of cities, emphasizing the integration of community voices and sustainable practices into urban development. Covering social learning, citizen engagement, and green policies in the digital age, while highlighting the importance of transparency in enhancing urban resilience. The role of technology in public safety, energy management, and healthcare is also examined, with innovative approaches presented for optimizing safety, comfort, and sustainability. Additionally, urban design and infrastructure and reimagined, by addressing how cities can adapt to evolving healthcare needs and future uncertainties while integrating Building Information Modelling and Life Cycle Assessment for more sustainable building practices.

In bringing together these varied yet interconnected themes, this book provides a comprehensive overview of the opportunities and challenges that lie ahead for urban environments. It is our hope that the ideas and solutions presented within these pages will inspire architects, policymakers, urban planners, and technologists to collaborate in crafting cities that not only meet the demands of the future but also enhance the quality of life for all their inhabitants.

As we stand on the brink of this exciting new era, we invite you to explore the possibilities and potentials of swarm intelligence in shaping the cities of the future.

August, 2024

The editors,
Leonor Domingos
Maria José Sousa

Contents

List of Contributors

Andrea Brambilla
Design & Health Lab, Department of Architecture, Built Environment, and Construction Engineering, Politecnico di Milano
Andreia de Bem Machado
Federal University of Santa Catarina, Brazil.
António Sacavém
Universidade Europeia, Portugal.
Bruno Mataloto
ISTAR – Information Sciences and Technologies and Architecture Research Center, ISCTE-IUL – University Institute of Lisbon.
Carlos Coutinho
Department of Information Science and Technology, ISCTE – University Institute of Lisbon
Davide Calandra
University of Turin, Department of Management "Valter Cantino".
Diogo da Silva
Associação Visionarium.
Erica Isa Mosca
Design & Health Lab, Department of Architecture, Built Environment, and Construction Engineering, Politecnico di Milano
Gabriel Osório de Barros
Director for Economic Analysis at the Office for Strategy and Studies (GEE) of the Portuguese Ministry of Economy and Maritime Affairs.
Ginevra Degregori
University of Turin, Department of Management "Valter Cantino".
Giovanni Borsoi
CERIS, Civil Engineering Research and Innovation for Sustainability, Instituto Superior Técnico, University of Lisbon.
Helena Belchior Rocha
Department of Political Science and Public Policy. ISCTE – University Institute of Lisbon
João Carlos Ferreira
Department of Applied Digital Technologies. ISCTE – University Institute of Lisbon

João Poças Martins
CONSTRUCT—GEQUALTEC, Faculty of Engineering (FEUP), University of Porto.
João Rodrigues dos Santos
ISEC Lisboa – Instituto Superior de Educação e Ciências, Portugal.
Leonor Domingos
Department of Applied Digital Technologies, ISCTE – University Institute of Lisbon.
Manuel Pio Silva
CNET – Centre New Energy Technologies.
Maria José Sousa
Department of Political Science and Public Policy ISCTE – University Institute of Lisbon.
Mohamed Nour El-Din
CONSTRUCT—GEQUALTEC, Faculty of Engineering (FEUP), University of Porto.
Nuno M.M. Ramos
CONSTRUCT—GEQUALTEC, Faculty of Engineering (FEUP), University of Porto.
Pedro F. Pereira
CONSTRUCT—GEQUALTEC, Faculty of Engineering (FEUP), University of Porto.
Rafael Rodrigues
CNET – Centre New Energy Technologies
Ricardo Resende
ISTAR-Information Sciences and Technologies and Architecture Research Center, ISCTE – University Institute of Lisbon, Department of Applied Digital Technologies.
Sara Parece
ISTAR-Information Sciences and Technologies and Architecture Research Center, ISCTE – University Institute of Lisbon, School of Technology and Architecture.
Silvana Secinaro
University of Turin, Department of Management "Valter Cantino".
Stefania Stellacci
ISTAR – Information Sciences and Technologies and Architecture Research Center, ISCTE-IUL – University Institute of Lisbon.
Stefano Capolongo
Design & Health Lab, Department of Architecture, Built Environment, and Construction Engineering, Politecnico di Milano.
Tianzhi Sun
Design & Health Lab, Department of Architecture, Built Environment, and Construction Engineering, Politecnico di Milano.
Vasco Rato
ISTAR-Information Sciences and Technologies and Architecture Research Center, ISCTE – University Institute of Lisbon, School of Technology and Architecture.

1 Social Learning and Citizens' Engagement in Future Cities

Andreia de Bem Machado,[1*] Gabriel Osório de Barros,[2]
João Rodrigues dos Santos[3] and António Sacavém[4]

1. Introduction

The initial decades of the 21st century have seen a significant rise in urbanization, with an estimated 68% of the global population now living in cities (United Nations, 2018). This shift necessitates advanced, sustainable solutions to manage urban complexity, for which Swarm Intelligence offers a compelling strategy, leveraging collective intelligence for problem-solving (Bonabeau et al., 1999). This chapter delves into Social Learning and Citizen Engagement, critical elements in this approach. Social Learning involves learning from peer interactions within a social framework, a key to spreading knowledge and fostering community-level innovation for sustainable city living (Bandura, 1977; Wenger, 1998). Citizen Engagement denotes the active participation of citizens in governance, essential for creating inclusive and responsive urban environments (Cornwall, 2008; Irvin and Stansbury, 2004). Acknowledging that top-down methods fall short for modern urban challenges, this chapter emphasizes that the future of urban planning lies in bottom-up, citizen-powered strategies that are sustainable and resilient (Benkler, 2006). It aims to showcase how Social Learning and Citizen Engagement, within the context of Swarm Intelligence, can cultivate cities that are more aligned with their residents' needs, promoting efficiency, sustainability, and a stronger community spirit (Bonabeau et al., 1999).

2. Swarm Intelligence: A Key to Future Cities

Swarm Intelligence is an emerging field that utilizes principles derived from the collective behavior of decentralized, self-organized systems to solve complex

[1] Federal University of Santa Catarina, Brazil.
[2] Director for Economic Analysis at the Office for Strategy and Studies (GEE) of the Portuguese Ministry of Economy and Maritime Affairs.
[3] ISEC Lisboa – Instituto Superior de Educação e Ciências, Portugal.
[4] Universidade Europeia, Portugal.
Email : gabriel.barros@gee.gov.pt; jcaopaulo.santos@iseclisboa.pt; antonio. sacavem@universidadeeuropeia.pt
* Corresponding author: andreia.bem@ufsc.br

problems (Bonabeau et al., 1999). The application of Swarm Intelligence to the urban context provides innovative ways to address complex issues inherent in rapidly evolving urban systems.

Swarm-based principles can be used to optimize various facets of urban living, from traffic management and infrastructure planning to environmental conservation and resource management. For instance, the use of Swarm Intelligence in traffic management systems can optimize traffic flow, reducing congestion and travel times (Dorigo and Stützle, 2004). Similarly, swarm-based solutions can enable more efficient use of resources by coordinating usage based on demand patterns across the city (Di Caro et al., 2005).

At the heart of Swarm Intelligence lies the principle of simple agents interacting to generate complex, efficient, and adaptive solutions (Bonabeau et al., 1999). This aligns closely with the concepts of Social Learning and Citizen Engagement, the key for future city development.

- **Facilitating Social Learning:** As cities grow increasingly interconnected, the need for effective Social Learning mechanisms becomes paramount. Social Learning, in the context of Swarm Intelligence, involves the exchange of knowledge and experiences between individuals, allowing for collective problem-solving and the diffusion of innovative solutions (Wenger, 1998). Similar to how birds learn to flock or fish to school, humans in urban environments can learn and adapt behaviors based on the collective intelligence of the group. Such social learning processes can enhance the efficiency, adaptability and resilience of urban systems, contributing to the overall livability of future cities.
- **Promoting Citizen Engagement:** Active citizen engagement is a linchpin for the successful application of Swarm Intelligence principles in urban settings. The more citizens actively participate and share their data and feedback, the more accurate and effective the swarm-based solutions become. Citizen engagement can take many forms, from contributing data through smart devices and Internet of Things (IoT) networks to participating in decision-making processes about urban development (Irvin and Stansbury, 2004). Active citizen engagement not only improves the effectiveness of swarm-based solutions but also fosters a sense of community and shared ownership of urban spaces.
- **Creating Responsive Cities:** Through Social Learning and Citizen Engagement, Swarm Intelligence can facilitate the development of responsive cities, which are characterized by their adaptability to the needs and behaviors of their citizens (Batty, 2013). By constantly learning from the behavior of its inhabitants, akin to the principles of Swarm Intelligence, a city can dynamically adapt its services, infrastructure, and policies (Cardullo and Kitchin, 2019). This constant adaptation and learning can lead to the creation of 'smart' cities - urban environments that leverage digital and telecommunication technologies to become more efficient, sustainable, and in-tune with their inhabitants' needs (Caragliu, Del Bo, and Nijkamp, 2011; Giffinger and Gudrun, 2010).

Consequently, this results in urban spaces that are not just designed for people, but dynamically evolve with them (Townsend, 2013).

In summary, Swarm Intelligence provides an innovative framework for managing the complexities of future cities. By integrating the principles of Social Learning and Citizen Engagement, swarm-based solutions can tap into the collective intelligence and adaptive capacity of urban dwellers.

3. Essential Concepts and Applications of Swarm Intelligence in Social Learning

Swarm Intelligence plays a pivotal role in social learning, leveraging the collective behavior of agents for interaction and knowledge sharing, as posited by Natarajan (2018). It draws inspiration from natural systems, such as bird predation strategies, to create algorithms applicable to human societies (Zhang and Qiu, 2020; Trelea, 2003). Key concepts of Swarm Intelligence in social learning include observation and imitation, where agents mimic successful behaviors, facilitating learning through imitation, with entropic forces playing a significant role (Tao, 2018; Wissner-Gross and Freer, 2013).

Social influence and information sharing are also integral, enabling agents to affect each other's behaviors and share knowledge, leading to collective decision-making and adaptive social behavior (Yang, 2023). This decentralized and adaptive organization enables complex behavioral patterns and systematic adjustment without a central authority (Acemoglu and Ozdaglar, 2011).

In collaborative learning, particularly through Internet of Medical Things (IoMT) and online learning communities, Swarm Intelligence promotes knowledge exchange and problem-solving, contributing to social skills improvement (Cheng et al., 2023; Tomar, 2017; Halepoto, 2017; Ramzan et al., 2022). Moreover, Swarm Intelligence fosters diversity and inclusion in social learning by encouraging interaction among diverse agents.

Personalized recommendation systems, such as those used by YouTube or Netflix, utilize Swarm Intelligence techniques to tailor content to user preferences (Tao, 2018; Bonabeau and Meyer, 2013). Adaptive learning platforms like Knewton and Peer-Predict leverage are Swarm Intelligence tools for customized learning experiences and collaborative knowledge exchange.

In social networks and online games, Swarm Intelligence techniques aid in collaborative learning and scientific advancements, as seen in Foldit and Stack Exchange, where users collectively solve puzzles or rank the best answers (Shi, 2015; Jeevana et al., 2020; Cheng, 2016). The Artificial Bee Colony algorithm exemplifies the application of Swarm Intelligence in algorithmic solutions for user engagement (Önder et al., 2014).

Swarm Intelligence's transformative potential in Social Learning is immense, shaping the social fabric through decentralized, self-organizing principles that enhance user experiences and inform decision-making. The next section aims to explore case studies demonstrating Swarm Intelligence's successful application in social learning, underscoring its effectiveness in creating adaptive, responsive, and inclusive communities.

4. Presentation of Relevant Case Studies

There are several examples of research projects in this area. Miao et al. (2023), for instance, in their study, "applied ten swarm intelligence-based algorithms to investigate their potential for energy-saving optimization control in an HVAC [heating, ventilation, and air-conditioning] system".

In a completely different area of research, according to Wang et al. (2013), "task allocation is a key issue of agent cooperation mechanism in Multi-Agent Systems". The case study *Ant Colony Optimization for Task Allocation*, initially presented by Marco Dorigo, in 1992, as part of his doctoral thesis (Colorni et al., 1991), explores how ant colony optimization algorithms can be applied to optimize task allocation in a social learning scenario. It investigates how a swarm of artificial ants can collectively learn and adapt their behavior to efficiently allocate tasks among themselves.

Another example of a very common study area within the scope of swarm intelligence applied to social learning is *Bird Flocking Behavior and Consensus Decision Making*. This study area examines the flocking behavior of birds and how they collectively make consensus decisions, such as selecting a new migration route. One of the case study examples in this area is *Collective Decision-Making in Homing Pigeons: Larger Flocks Take Longer to Decide but Do Not Make Better Decisions* (Santos, et al., 2017). In the last paragraph of the study, it is possible to read: "Pigeon flock decision-making is overall a remarkable phenomenon, illustrating the complexity of collective animal movement." By studying birds' social learning mechanisms, researchers gain insights into how to design algorithms for collective decision-making in swarm intelligence systems.

Another important case study in the scope of this chapter, for example, is "*What Can Honeybees Tell Us About Social Learning*?, developed by Clark and Kimbrough (2020). The study focuses "on aspects of the model pertaining both to imitation and social learning and to the model as kind of metaheuristic". Honeybee swarms exhibit sophisticated social learning behavior, where individual bees share information about the location of new food sources through dances. This assumes particular interest for research around swarm intelligence.

One last example of a very interesting research area within the scope of the theme under analysis in this chapter is *Opinion Dynamics in Social Networks*. This research area focuses on the dynamics of opinion formation and diffusion in social networks. In this regard, Acemoglu and Ozdaglar (2011) conclude that "social networks play an important role in opinion formation in practice [simply] because we tend to talk to people in our social network". By analyzing how opinions spread and evolve within a network of individuals, researchers gain insights into collective decision-making, information cascades, and the emergence of social consensus.

All the previous case study examples highlight the application of swarm intelligence principles to social learning scenarios, providing valuable insights into collective decision making, task allocation, information sharing, and opinion dynamics within a swarm. Ultimately, they all revolve around one of the key aspects of swarm intelligence: the algorithmic calibration needed to improve social learning systems.

This section provided a comprehensive exploration of a range of research projects that exemplify the application of swarm intelligence in social learning scenarios. These findings collectively highlight the potential of swarm intelligence in facilitating social learning, whether through energy conservation, task allocation, decision-making, or information sharing. Each case study has helped deepen our understanding of the diverse applications of swarm intelligence, demonstrating its adaptability and utility across a broad range of contexts.

Moving forward, we turn our attention to the more social aspect of swarm intelligence: its potential to transform citizen engagement in various scenarios, from local community decisions to disaster response. In the next section, we delve into how the principles and applications of swarm intelligence can foster an active, empowered, and engaged citizenry, reflecting on a range of scenarios such as policymaking, crisis management, and smart city initiatives. Through this exploration, we will broaden our understanding of how swarm intelligence can be harnessed to create more inclusive, effective, and resilient social systems.

5. Infusing Citizen Engagement

Swarm intelligence, inspired by social insects, is highlighted to enhance citizen engagement and collective decision-making. This concept allows groups, like schools of fish or ant colonies, to surpass individual knowledge when responding to challenges, as noted by Ioannou (2017). Tang et al. (2021) emphasize its advantages in adaptability and performance, while Schranz et al. (2021) explore human interaction with swarms, and Metcalf et al. (2019) suggest it allows for merging individual insights in real-time, aiding in decision-making.

The application of swarm intelligence is broad, including disaster response in smart cities, where citizens contribute like a 'swarm' to crisis management, demonstrating collective problem-solving as seen in the work of Palen et al. (2010). Rosenberg (2015) argues that this not only speeds up consensus but also fosters unity among participants.

Swarm intelligence can also facilitate operational scenarios, such as guiding drone swarms in disaster relief, where human influence can optimize the swarm's effectiveness, a concept explored by Rodriguez and Hilaire (2023). Zedadra et al. (2019) and Alfeo et al. (2019) discuss its role in smart cities, where citizens actively engage with urban infrastructure, contributing data that inform decision-making and resource management.

A practical case is the Boston Marathon bombing, where swarm intelligence was evident in the public's instinctive collaboration, a study of which was undertaken by Marcus et al. (2019) from the Harvard National Preparedness Leadership Initiative.

In summary, swarm intelligence offers a paradigm where collective behavior and input can lead to more effective decision-making and problem-solving, with potential applications in community participation, smart city development, and emergency responses. The theory, supported by various studies, suggests an approach to harness the collective power of citizens, leading to inclusive and resilient social systems. The next step is a bibliometric study to deepen the understanding of swarm intelligence literature.

The aim of the next section is to quantify the body of research dedicated to the intersection of social learning, citizen engagement, and future cities, using a bibliometric study. Bibliometrics is a field that utilizes mathematical and statistical methods to analyze and map documents and citations within a specific area of study. By leveraging this tool, we can explore the research landscape in a more organized and systematic manner, identifying key contributors, emerging trends, and important publications in the field.

6. Methodology

The bibliometric analysis conducted through the Web of Science database focused on scientific literature concerning social learning and citizen involvement in future cities. This study was structured into three main stages: planning, data collection, and bibliometric analysis. Bibliometrix, compatible with the Web of Science and described as a comprehensive RStudio package for quantitative research by Aria and Cuccurullo (2017), facilitated the bibliometric assessment. This approach allowed for the organization of crucial information such as temporal distribution, key authors and institutions, as well as the frequency and impact of publications in the field (Bibliometrix, 2022). Linnenluecke (2019) emphasized that Bibliometrics applies mathematical and statistical methods to analyze bibliographic records, aiding in a systematic representation of research outputs and minimizing biases. The study, following an exploratory-descriptive classification, aimed to elucidate the state of research on the topic, with a systematic search executed in July 2023, without language or temporal limitations. It included only 'article' type documents, resulting in 138 indexed papers from 1998 to 2023, and it underlined the significance of clear search term definitions in the planning phase.

7. Bibliometrics Results

As a result of this collection, it was identified that the 138 documents identified in the search were written by 437 authors, linked to 41 different countries. A total of 551 keywords were used. Table 1 presents the result of this data collection in a general bibliometric analysis.

Table 1 Bibliometric data.

Description	*Results*
MAIN INFORMATION ABOUT DATA	
Timespan	1998: 2023
Sources (Journals, Books, etc.)	109
Documents	138
Annual Growth Rate %	7,43
DOCUMENT CONTENTS	
Keywords Plus (ID)	433
Author's Keywords (DE)	551
AUTHORS	
Authors	413
Authors of single-authored docs	24
DOCUMENT TYPES	
Articles	113
Articles; early access	3
Articles; proceedings paper	3
Proceedings papers	15
Reviews	4

The eligible articles in the Web of Science database were published from 1998–2023. In 1998, only one article was published. There was a significant increase in publications from the year 2016, in that year, 12 documents were published.

Of the 138 papers, there is a varied list of authors, institutions, and countries that stand out in the research on Social Learning and citizens engagement in future cities. When analyzing the 20 countries with the highest number of citations in the area, it can be seen that the USA stands out with 36% of the total citations, a total of 144 citations. In the second place, the United Kingdom stands out with 13% of the citations.

Figure 1 presents the intensity of publication by country and the relationship established between them, through citations between published works, the continents: South America, North America, Europe and Asia, as shown in the following image.

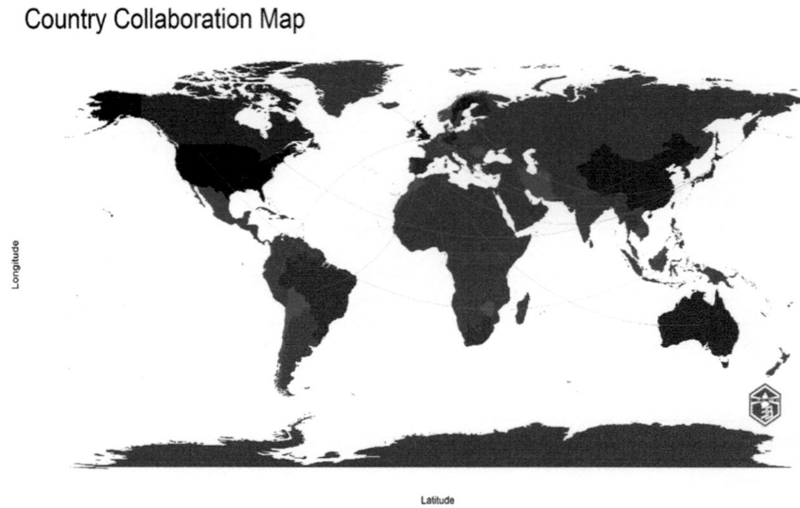

Fig. 1 Spatial distribution and relationships of publications on the study topic.

Another analysis is related to the identification of authors. The author who has the most production on the subject is Marc Wolfram of the Leibniz Institut fur okologische Raumentwicklung with 5 publications in the area, as shown in Figure 2.

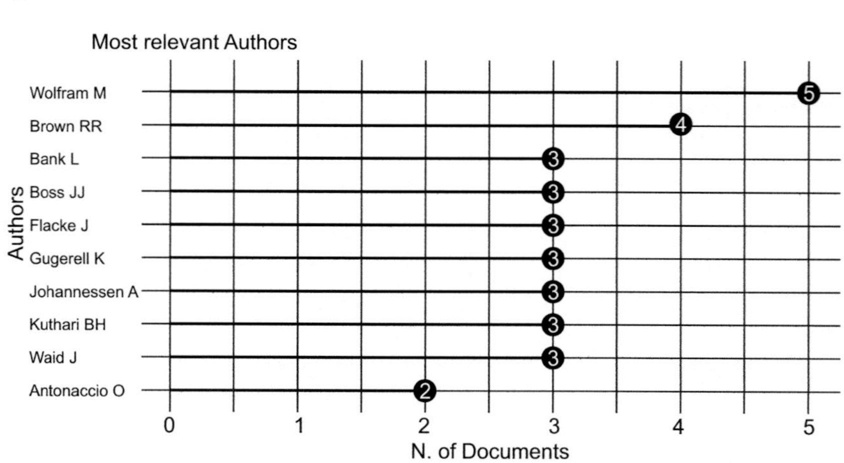

Fig. 2 Authors with the highest number of publications on the search topic.

The productivity and relevance of the production over time of the main authors are represented in Figure 3, where the size of the point represents the number of publications, and the intensity of the color, the number of citations of the annual publications. It is observed that the productivity and relevance of publications over time are dynamic.

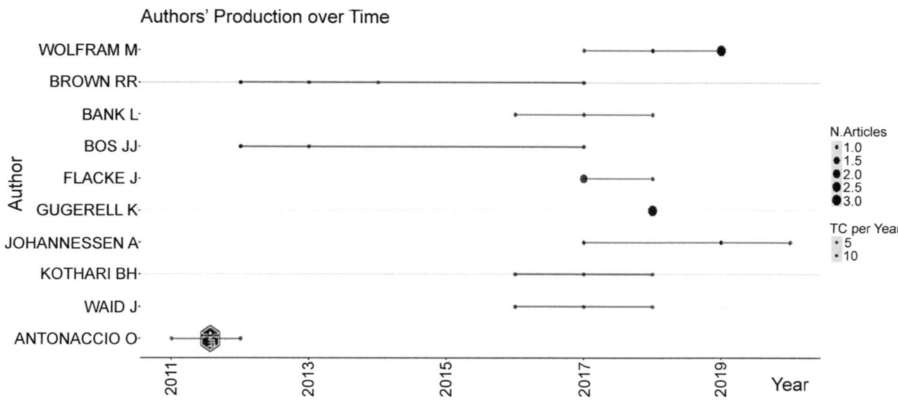

Fig. 3 Productivity and relevance of publications by author.

The documents analyzed were published in 109 different journals, and among the total of 138 studies, 8 (25%) were published in a single journal *Sustainability*, as shown in Figure 4, which presents the 10 scientific sources with the most publications on the theme of this study.

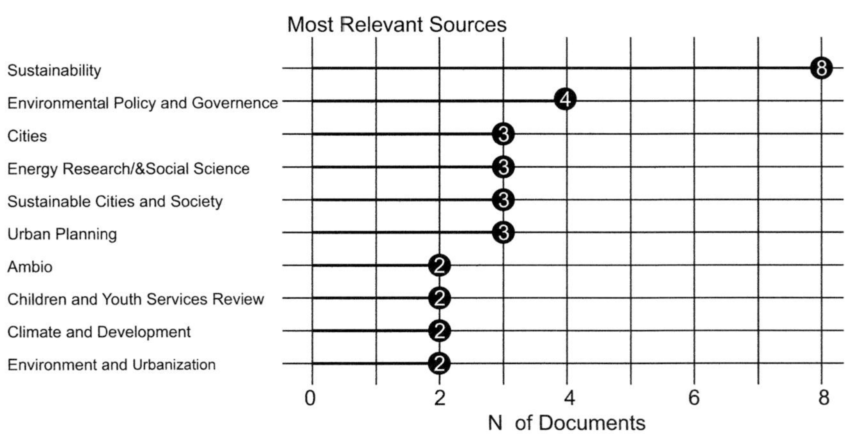

Fig. 4 Scientific sources with the most publications.

From the bibliometric analysis, based on the retrieved work group, in the base of 551 keywords indicated by the authors, cities gained prominence with 29 occurrences.

'Governance' is also highlighted, from which we conclude that it is an area of application in social learning and citizen engagement in future cities explored by the literature.

8. Challenges and Opportunities

Currently, scientific research in this area aims to help find answers to the current and future major challenges facing the application of swarm intelligence to social learning.

One of these major challenges is to manage, in all dimensions, the engagement of citizens in swarm intelligence processes. If, on the one hand, it is essential for the processes to collect and share relevant data on behaviors, opinions, and social interactions, on the other hand, there are also concerns about privacy and data security, as well as technical challenges to deal with the diversity and quality of the data collected. In fact, the reliability and bias of the data collected to apply swarm intelligence effectively is one of the biggest challenges facing the present and the future. In fact, Kroll et al. (2016), referring to Schwartz (1992), reminds that "the design and implementation of automated decision systems can be vulnerable to a variety of problems that can result in systematically faulty and biased determinations". While data, in order to be reliable, should be representative of the diversity of perspectives and experiences of engaged citizens, biases can occur, either due to unequal participation of citizens or due to polarization of opinions, which can affect the effectiveness and objectivity of swarm intelligence.

To counter the possibility of bias, the algorithms developed must be increasingly robust. Algorithms must be increasingly accurate at processing large amounts of data, identifying relevant patterns, and generating meaningful recommendations or insights to support social learning and civic engagement. This is also a present concern and will remain so in the future.

On the subject of civic engagement, one limitation that persists as significant today is the digital inclusion deficit. In this regard, the work *Digital Skill Sets for Diverse Users: A Comparison Framework for Curriculum and Competencies* (Wedlake et al., 2019) is highlighted. It is "a tool for funders, governments, and training providers to help focus and improve their digital inclusion efforts". In some parts of the globe, digital accessibility and inclusion are still not desirable. Not all citizens have equal access to digital technologies or the skills to actively participate. Ensuring that tools and platforms are accessible and easy to use is the key to promoting broad and inclusive participation, and therefore effective swarm intelligence integration.

Finally, it is also worth noting the challenge that the increasing possibility of data manipulation and disinformation practices poses now and in the future for the effectiveness of swarm intelligence and its application to social learning processes. Kroll et al. (2016) is ironic about this matter, noting that "it should not be surprising that [...] computational processes require some redesign of the processes themselves [because] the most accountable processes are those that are designed with accountability in mind".

Malicious actors may attempt to influence swarm intelligence results through data manipulation, propagation of false information or fraudulent engagement. Technological prevention and the detection and mitigation of such practices are essential to maintain the integrity of the process. Similarly, the application of swarm intelligence also requires adequate transparency and governance. In this regard, it is important to retain Yang's (2023) idea:

Swarm system [...] to model the intelligent agents' motivation, present corresponding actions and tactics adapting to various situations, and optimize their behaviors and strategies through learning from interactions with environments and previous experience.

It is essential to ensure the transparency of data collection and use processes, as well as the explainability of the algorithms employed. Taking care of these concerns will ensure the reliability of the results and an increasing involvement of citizens.

Thus, addressing the challenges outlined above requires a careful approach. Ethical, technical, and social aspects must therefore be considered, ensuring that integration is beneficial and inclusive for society as a whole.

Therefore, any set of solutions that aim to address the current and future challenges in integrating swarm intelligence logic into social learning processes must necessarily include the strengthening and enhancement of the inherent regulatory framework. With the speed at which technology evolves (see the case of artificial intelligence), what Kroll et al. (2016) already warned in 2016 is more valid today than ever: "The accountability mechanisms and legal standards that govern decision processes have not kept pace with technology".

Improved policy and regulation can help ensure transparency, accountability, and protection of citizens' rights in the context of increasing swarm intelligence integration. Data protection laws, ethical guidelines, and adequate oversight mechanisms are some of the aspects that should be considered.

Equally important will be the development of algorithms that take into account fairness, impartiality, and bias detection. In this context, Chesney and Citron (2019) draw attention to "the amplifying power of social media and the confounding influence of cognitive biases". For example, transparency of algorithms and even peer review can be incorporated to ensure that swarm intelligence results are reliable and fair.

From another perspective, overcoming disparities in access to technology should also be a focus in the present. To emphasize the importance of access to technology today, we highlight what Ramzan et al. (2022) said in the context of the IoT and cloud-based systems: "These systems are able to transfer the enormous data packages that are assembled by IoT through the internet." It is important to invest in initiatives that promote education and digital inclusion. This can include programs for access to infrastructure and connectivity, digital literacy, and promotion of digital skills to ensure that all citizens have the opportunity to participate in the process of social learning and civic engagement. Digital literacy is also a prerequisite to ensure more education and awareness on disinformation. To this end, the internet is an indispensable tool. Natarajan (2018) poses this question: "Is Internet the evolved version of the Noosphere?" This idea was firstly introduced by the paleontologist and Jesuit priest Pierre Teilhard de Chardin (1881–1955) in his book *Le phénomène humain* "More specifically, it represents the stage of evolutionary development dominated by consciousness, the mind, and interpersonal relationships" (Natarajan, 2018). This reflection is truly suggestive of the importance of the WWW today.

Measures and initiatives aimed at increasing awareness and education on disinformation and data manipulation should include the development of critical digital literacy skills, information evaluation, and source verification, all to empower citizens to protect themselves against the dissemination of misleading information.

On the other hand, "innovation scholars agree that the bridging of heterogeneous networks and ideas can accelerate innovation" (Toivonen, 2013). In this regard, ensuring the maximum effectiveness of the above solutions requires a collaborative approach. Stakeholders, including governments, educational institutions, ethicists, researchers, and civil society, should work together to develop guidelines, share best practices, and promote discussions on the impacts and implications of integrating social learning practices, and related citizen engagement, with the philosophy behind swarm intelligence.

Finally, exploring the potential of swarm intelligence in urban planning and citizen engagement necessitates acknowledging its limitations and risks. Despite offering innovative collective decision-making and problem-solving solutions, swarm intelligence's applicability in complex urban settings faces significant challenges. Scalability issues may arise as cities expand in complexity, potentially surpassing the algorithms' data management and decision-making capabilities. Furthermore, swarm intelligence models, based on simplified behavioral assumptions, might not fully capture the intricate human behaviors and social dynamics of urban environments. Technical obstacles, including the need for precise parameter tuning and algorithm vulnerability to disturbances, complicate their use. Ethical concerns about privacy, fairness, and transparency, coupled with the risk of diminishing human oversight in planning, call for a balanced approach. Additionally, implementing swarm intelligence systems introduces security risks like susceptibility to cyberattacks or malicious manipulations. These challenges underscore the need for continued research and interdisciplinary efforts to address the ethical and technical intricacies of incorporating swarm intelligence in urban development.

Despite of the myriad challenges presented, it is also crucial to understand that these obstacles serve as steppingstones towards a future replete with untapped potential. The marriage of social learning and swarm intelligence provides a unique platform to revolutionize citizen engagement and how societies collectively learn, adapt, and grow. With this perspective, let's now turn our attention to the future opportunities that the integration of social learning and citizen engagement with swarm intelligence can offer.

9. Future Opportunities in Integrating Social Learning and Citizen Engagement with Swarm Intelligence

Notwithstanding the challenges, the integration of social learning, and the necessary engagement of citizens in the process, with the logic of swarm intelligence, also presents several promising opportunities. Tao (2018), drawing on Bonabeau (1999, 2001), Woolley et al. (2010), and Muchnik and Taylor (2013), highlights that "due to the fascinating prospect of Swarm Intelligence, it has been applied to many fields, such as artificial intelligence, sociology, business management and so on".

One of the great opportunities arising from the application of swarm intelligence to social learning is the improvement of decision-making processes. The combination of diverse perspectives and knowledge of the citizens involved, together with algorithms created in the context of swarm intelligence, can result in more informed decisions, built collectively and thus more aligned with the needs of society as a whole. Zhang and Qiu (2020), in this regard, highlight that "based on the logic of natural selection, the approaches of swarm intelligence algorithm mimics individual and in-group behaviors of species to seek the optimal solution". And, regarding the reinforcement of collectively made decisions, the second great opportunity to result from the union of swarm intelligence with social learning arises: the expansion of citizen participation. Indeed, the integration of swarm intelligence with social learning allows a larger number of people to contribute their ideas, knowledge, and opinions. This can promote greater inclusion and diversity of perspectives in the civic engagement process, resulting in more representative and legitimized decisions by the community.

Another very relevant opportunity is the enhancement of the personalization of learning and the increase of the adaptive dimension of the process. In this regard, attention is drawn to a study by Yang and Parasuraman (2020) about a group of swarm robots cooperating to complete some tasks. "Since the tasks are dynamically assigned, the robots need to change their plans and adapt to the new scenario to guarantee the group utility." In this sense, by analyzing the behavior and preferences of the citizens involved, swarm intelligence algorithms can provide personalized learning recommendations, adapted to the individual needs and pace of each person. As learning processes become more effective, we are likely to see a boost in social innovation. By bringing together the collective ideas and knowledge of engaged citizens, combined with the power of swarm intelligence algorithms, innovative solutions to complex social challenges can be found, driving positive change in society.

In conclusion, the integration of social learning, citizen engagement, and swarm intelligence offers promising avenues for shaping the future of societal decision-making and learning processes. From enhancing the efficacy of decision-making through collective wisdom to broadening the scope of citizen participation, these techniques hold the promise of crafting a more inclusive, diverse, and representative civic landscape. Furthermore, they herald the potential for truly personalized learning experiences, adapting to each individual's pace and preferences, fostering a more engaged, informed, and participatory citizenry. As a stepping-stone for social innovation, this integration opens doors to devising creative solutions to complex societal challenges, acting as a catalyst for positive change in our communities.

While we acknowledge the potential of these opportunities, it is critical to keep in mind the challenges we outlined earlier. Navigating these challenges requires a balanced and multi-faceted approach, considering ethical, technical, and social considerations, ensuring that the integration of these concepts is beneficial to society at large.

Conclusion

Throughout this discussion, we have ventured into the exploration of the intersection of social learning, citizen engagement, and swarm intelligence, specifically in the context of future cities. We are prompted by the recognition that urban environments of the future should be more adaptable, efficient, and responsive to the needs of their residents. The most effective way to understand and respond to these needs is through harnessing the collective intelligence of the citizens themselves.

The purpose of our exploration was to illuminate the theory and practice of applying swarm intelligence to enhance social learning and civic participation, while identifying and addressing the associated challenges and opportunities. This exploration was guided by an extensive literature review and enriched by relevant case studies.

Our analysis reveals how swarm intelligence, underpinned by principles of decentralized decision-making and collective behavior, can significantly amplify the inclusivity, diversity, and adaptability of social learning and citizen engagement processes. It demonstrates the potential of swarm intelligence to revolutionize decision-making, give voice to diverse perspectives, cultivate personalized learning experiences, and empower citizens to collaboratively address complex challenges.

However, the integration of swarm intelligence into social learning processes and civic engagement does not come without its obstacles. Issues around data privacy and security, digital inclusivity, potential biases, and susceptibility to data manipulation emerge as significant challenges, such as personal data vulnerability, surveillance concerns, data breach consequences, access disparities, skill gaps, content and language barriers, algorithmic bias, information echo chambers, and manipulation and fake news. There is also a need for robust and interpretable algorithms that can effectively handle vast and diverse data streams. Addressing these challenges requires a concerted effort from various stakeholders, innovative solutions, and robust regulation and oversight mechanisms.

Despite these challenges, the opportunities are vast. Integrating swarm intelligence into social learning and citizen engagement could democratize decision-making processes, foster collective problem-solving and innovation, and make urban environments more adaptable, efficient, and responsive to the needs of their inhabitants. This illustrates the real potential of smart and future cities.

In conclusion, Swarm Intelligence provides a powerful framework for harnessing the collective intelligence and engagement of citizens in future cities, contributing to creating urban environments that are more adaptable, efficient, and responsive to the needs of their inhabitants. As we navigate the complexities of the digital revolution and the fourth industrial revolution, the integration of social learning, citizen engagement, and swarm intelligence emerge as a promising path forward.

Although the insights gained from this study are significant, further research is needed to unravel the complexities of these intertwined concepts and shed light on their real-world implications. Detailed case studies, practical examples, and wider geographical representation in the research space will further enrich our understanding. As we continue to envision and shape the cities of the future, this work offers a valuable perspective, one that champions collective intelligence, shared responsibility, and citizen engagement. It underscores the role of swarm intelligence in creating more inclusive, adaptable, and efficient urban environments. Thus, our journey to the future cities is not a solitary one but a collective endeavor, one that is guided by the principles of swarm intelligence and the power of collective decision-making.

References

Acemoglu, D. and Ozdaglar, A. (2011). Opinion dynamics and learning in social networks. *Dynamic Games and Applications*, *1*, 3–49.

Alfeo, A.L., Ferrer, E.C., Carrillo, Y.L., Grignard, A., Pastor, L.A., Sleeper, D.T., ... and Pentland, A. (2019). Urban Swarms: A new approach for autonomous waste management. *In*: *2019 International Conference on Robotics and Automation*. IEEE.

Alammary, A., Alhazmi, S., Almasri, M. and Gillani, S. (2019). Blockchain-based applications in education: A systematic review. *Applied Sciences*, *9*(12), 2400.

Aria, M. and Cuccurullo, C. (2017). Bibliometrix: An R-tool for comprehensive science mapping analysis. *Journal of Informetrics*, *11*(4), 959–75.

Bandura, A. and Walters, R.H. (1977). *Social Learning Theory* (Vol. 1). Prentice Hall: Englewood Cliffs.

Batty, M. (2013). *The New Science of Cities*. MIT Press.

Benkler, Y. (2006). *The Wealth of Networks: How Social Production Transforms Markets and Freedom*. Yale University Press.

Bibliometrix. (2022). https://www.bibliometrix.org/

Bonabeau, E., Dorigo, M. and Theraulaz, G. (1999). *Swarm Intelligence: From Natural to Artificial Systems* (No. 1). Oxford University Press.

Caragliu, A., Del Bo, C. and Nijkamp, P. (2011). Smart Cities in Europe. *Journal of Urban Technology*, *18*(0048).

Cardullo, P. and Kitchin, R. (2019). Being a 'citizen' in the smart city: Up and down the scaffold of smart citizen participation in Dublin, Ireland. *GeoJournal*, *84*(4), 1–13.

Cheng, Z., Liu, Y., Pan, Y., Zhao, L., Deng, X. and Zhu, C. (NA). Blockchain-Based Data Sharing for IoMT Networks: An Efficient and Privacy-Enhanced Collaborative Learning Approach with Joint Protection Strategy. *Computer Networks.*

Chesney, B. and Citron, D. (2019). Deep fakes: A looming challenge for privacy, democracy, and national security. *Calif. L. Rev., 107*, 1753.

Clark, R. and Kimbrough, S.O. (2020). What Can Honeybees Tell Us About Social Learning? *In: Proceedings of the 2018 Conference of the Computational Social Science Society of the Americas 1*. Springer International Publishing.

Colorni, A., Dorigo, M. and Maniezzo, V. (1991). Distributed Optimization by Ant Colonies, actes de la première conférence européenne sur la vie artificielle. *Elsevier Publishing.*

Cornwall, A. (2008). Unpacking 'Participation': Models, meanings, and practices. *Community Development Journal, 43*(3), 269–83.

Di Caro, G., Ducatelle, F. and Gambardella, L.M. (2005). AntHocNet: An adaptive nature-inspired algorithm for routing in mobile ad hoc networks. *European Transactions on Telecommunications, 16*(5).

Dorigo, M. and Stützle, T. (2004). *Ant Colony Optimization.* MIT Press.

Giffinger, R. and Gudrun, H. (2010). Smart cities ranking: An effective instrument for the positioning of cities? *ACE: Architecture, City, and Environment, 4*(12).

Hamidi, H. and Kamankesh, A. (2018). An approach to intelligent traffic management system using a multi-agent system. *International Journal of Intelligent Transportation Systems Research, 16*(4).

Huang, S., Ye, J., Xu, Y., Li, J., Fu, S., Wang, Y. and Qin, Y. (2023). Ultralow confinement loss anti-resonant fibers designed by swarm intelligence. *Optics Communications, 530*, 129208.

Ioannou, C.C. (2017). Swarm intelligence in fish? The difficulty in demonstrating distributed and self-organised collective intelligence in (some) animal groups. *Behavioural Processes, 141*(2), 141–51.

Irvin, R.A. and Stansbury, J. (2004). Citizen Participation in Decision Making: Is It Worth the Effort? *Public Administration Review, 64*(1), 55–65.

Jeevana Jyothi, P., Raiyyan, S.M. and Rajasekhar, T. (2020). MSA: An Application of Brain Storm Optimization Algorithm. *In: 2nd International Conference on Communication & Information Processing*. Singapore, November 26–29, 2016.

Jennings, N.R., Moreau, L., Nicholson, D., Ramchurn, S., Roberts, S., Rodden, T. and Rogers, A. (2014). Human-agent collectives. *Communications of the ACM, 57*(12), 80–88.

Kroll, J.A., Huey, J., Barocas, S., Felten, E.W., Reidenberg, J.R., Robinson, D.G. and Yu, H. (2017). Accountable algorithms. *University of Pennsylvania Law Review, 165*, 633.

Laghari, A.A., He, H., Halepoto, I.A., Memon, M.S. and Parveen, S. (2017). Analysis of quality of experience frameworks for cloud computing. *IJCSNS, 17*(12), 228.

Linnenluecke, M.K. (2017). Resilience in business and management research: A review of influential publications and a research agenda: Resilience in business and management research. International *Journal of Management Reviews, 19*(1), 4–30.

Marcus, L.J., McNulty, E.J., Henderson, J.M. and Dorn, B.C. (2019). *You're it: Crisis, Change, and How to Lead when it Matters Most.* Hachett Book Group. https://www. publicaffairsbooks.com.

Metcalf, L., Askay, D.A. and Rosenberg, L.B. (2019). Keeping humans in the loop: Pooling knowledge through artificial swarm intelligence to improve business decision-making. *California Management Review, 61*(4), 84–109.

Meyer, C. and Bonabeau, E. (2001). Swarm intelligence: A whole new way to think about business. *Harvard Business Review, 79*(5), 106–114, 165.

Miao, Y., Yao, Y., Hong, X., Xiong, L., Zhang, F. and Chen, W. (2023). Research on optimal control of HVAC system using swarm intelligence algorithms. *Building and Environment, 241*, 110467.

Muchnik, L., Aral, S. and Taylor, S.J. (2013). Social influence bias: A randomized experiment. *Science, 341*(6146), 647–651.

Natarajan, S. (2018). Emergence. *SSRN.*

Önder, E., Ozdemir, M. and Yıldırım, B.F. (2013). Combinatorial Optimization Using Artificial Bee Colony Algorithm and Particle Swarm Optimization Supported Genetic Algorithm. *Kafkas University Journal of Economics and Administrative Sciences Faculty*, *4*(6).

Palen, L., Anderson, K.M., Mark, G., Martin, J., Sicker, D., Palmer, M. and Grunwald, D. (2010). A vision for technology-mediated support for public participation and assistance in mass emergencies & disasters. *In*: *ACM-BCS Visions of Computer Science 2010, Conference,* University of California.

Ramzan, M.S., Asghar, A., Ullah, A., Jhanji, N.Z., Alsolami, F. and Ahmad, I. (2022). Enhanced Artificial Bee Colony Based Optimization for Mitigating Replication in Large Data for Internet of Things (IoT). *Available at SSRN 4281666.*

Rodriguez, S. and Hilaire, V. (2023). A methodological approach for the analysis and design of Human–Swarm interactions based upon feedback loops. *Expert Systems with Applications*, *217*, 119482.

Rosenberg, L.B. (2015). Human Swarms, a real-time method for collective intelligence. *In*: *Artificial Life Conference Proceedings*. MIT Press.

Santos, C.D., Przybyzin, S., Wikelski, M. and Dechmann, D.K. (2016). Collective decision-making in homing pigeons: Larger flocks take longer to decide but do not make better decisions. *Plos One*, *11*(2), e0147497.

Schranz, M., Di Caro, G.A., Schmickl, T., Elmenreich, W., Arvin, F., Şekercioğlu, A. and Sende, M. (2021). Swarm intelligence and cyber-physical systems: Concepts, challenges, and future trends. *Swarm and Evolutionary Computation*, *60*, 100762.

Shi, Y. (2011). An Optimization Algorithm Based on Brainstorming Process. *International Journal of Swarm Intelligence Research*, *2*(4).

Starr, D. (2020). What should crisis leadership look like? *The New Yorker*.

Tang, J., Liu, G. and Pan, Q. (2021). A review on representative swarm intelligence algorithms for solving optimization problems: Applications and trends. *IEEE/CAA Journal of Automatica Sinica*, *8*(10).

Tao, Y. (2018). Social brain: A perspective of swarm intelligence in humans. *SSRN 3302461*.

Tao, Y. (2018). Swarm intelligence in humans: A perspective of emergent evolution. *Physica A: Statistical Mechanics and its Applications*, *502* (C), 436–46.

Toivonen, T. (2013). The Emergence of the Social Innovation Community: Towards Collaborative Changemaking? *SSRN 2369540*.

Tomar, A., Kumar, S., Ghanshala, M., Mishra, S. and Gupta, A. (2022). Optimizing Degree of Imbalance and Utilization Cost in Cloud Environment using Effective Particle Swarm Method. *SSRN 4293597*.

Townsend, A. (2013). *Smart Cities: Big Data, Civic Hackers, and the Quest for a New Utopia*. W.W. Norton & Company.

Trelea, I.C. (2003). The particle swarm optimization algorithm: Convergence analysis and parameter selection. *Information Processing Letters*, *85*(6).

United Nations. (2018). 68% of the world population projected to live in urban areas by 2050, says UN. https://www.un.org/development/desa/en/news/population/2018-revision-of-world-urbanization-prospects.html.

Wang, L., Wang, Z., Hu, S. and Liu, L. (2013). Ant colony optimization for task allocation in multi-agent systems. *China Communications*, *10*(3).

Wedlake, S., Keyes, D. and Lothian, K. (2019). Digital skill sets for diverse users: A comparison framework for curriculum and competencies. *In*: *TPRC47: The 47th Research Conference on Communication, Information and Internet Policy*. Washington, DC., September 20–21.

Wenger, E. (1998). *Communities of Practice: Learning, Meaning, and Identity*. Cambridge University Press.

Wissner-Gross, A.D. and Freer, C.E. (2013). Causal entropic forces. *Physical Review Letters*, *110*(16), 168702.

Woolley, A.W., Chabris, C.F., Pentland, A., Hashmi, N. and Malone, T.W. (2010). Evidence for a collective intelligence factor in the performance of human groups. *Science*, *330*(6004).

Woszezenki, C.R., Besen, F., Santos, J.L. and Steil, A.V. (2013). Desaprendizagem Organizacional: uma revisão bibliométrica e analítica da literatura. *Perspectivas em Gestão & Conhecimento*, *3*(3).

Yang, Q. and Parasuraman, R. (2020). Hierarchical needs-based self-adaptive framework for cooperative multi-robot system.: *2020 IEEE International Conference on Systems, Man, and Cybernetics*. Toronto, Canada, March.

Yang, Q. (2023). Hierarchical Needs-driven Self-adaptive Multi-agent Systems: From Individual Behaviors to Swarm Intelligence. *SSRN 4334527.*

Zedadra, O., Guerrieri, A., Jouandeau, N., Spezzano, G., Seridi, H. and Fortino, G. (2019). Swarm intelligence and IoT-based smart cities: A review. *In*: *The Internet of Things for Smart Urban Ecosystems,* 177–200. ResearchGate.

Zhang, R. and Qiu, Z. (2020). Optimizing hyper-parameters of neural networks with swarm intelligence: A novel framework for credit scoring. *Plos One*, *15*(6), e0234254.

2 Technology as a Social Learning and Citizen Engagement Facilitator

Diogo da Silva[1*] and Helena Belchior Rocha[2]

1. Social Learning and Citizen Engagement in Future Cities

As people, we get strongly attached to the places we live in. We want them to improve, adapt to our needs, evolve together with us. As of 2024, more than 55% of human population lives in cities (Baeumler et al., 2021) and is concerned about not only their present issues, but also future ones, as these cities will be home to future generations.

One of the key aspects brought up whenever the concept of Future Cities is discussed is the role of technology. The rapid pace at which it has evolved, thanks to the work of Federico Faggin, Marcian 'Ted' Hoff Jr. and Stanley Mazor in creating the microprocessor (Faggin, Hoff, Mazor, and Shima, 1996) and the consequential advances, allowed for information to be shared, analysed, reproduced, and reworked.

The science-fiction of old is future in the present and many key technologies will play a part in citizen participation and social learning. Some of the examples include Artificial Intelligence, advanced communication interfaces (such as ocular enhancement devices) for the Metaverse, and Game-based engagement. These advancements are not a mere dream of visionaries but have already proven their role in the modern world.

As recent history shows, society is also constantly evolving, often at an unplanned pace, due to unforeseen circumstances. In these moments, as in the case of the SARS-CoV-2 pandemic or the 9/11 attacks on World Trade Center (Kornbluh, 2003), citizen learning and engagement have been crucial for the benefit of all of society. A terrorist attack initiated the biggest changes in how we operate security everywhere and its impact is still around, whilst the lockdown showed many that their job doesn't require them to commute and engage in person with others and, for some, isolation went from a struggle to a common way of living. However, for many these have been times of uncertainty and misinformation leading to damages (societal and individual, with special regards to mental health), the effects of which are lasting till today. Moreover, they served as an example when social learning and exchange of information has proven crucial in order to understand how to

[1] Associação Visionarium
[2] Iscte – Instituto Universitário de Lisboa
Email : h_rocha@iscte-iul.pt
* Corresponding author: contact@diogodasilva.eu

react and operate when in presence of elevated threat levels and how to contain the spread of a virus and design treatment and prevention methods. Today, the citizens' collective memory of those events can confirm how important the technology of information has been to decide the future course of the pandemic management or terrorist attack prevention. It relied on full citizen engagement in the greater cause of health and safety of others, which could only be achieved if the threat is understood and the good practice of maintaining distance and usage of personal protection sources are internalized and adopted. Let it serve as a reminder for the citizens and authorities of the future smart cities to value social learning importance not only in risk management but in the day-to-day practice of internal policy.

2. Understanding Social Learning and Citizen Engagement

Before initiating any discussion, it is imperative to firstly present what Social Learning Theory and Citizen Engagement encompass as well as how they associate between one another.

3. Social Learning and the Social Learning Theory

To address the complexities that arise from learning and interacting within the current and future world, it's required that we establish a baseline of understanding of what Social Learning and the Social Learning Theory (SL) introduced by Bandura (1977) represent.

The concept of SL can be considered an intermediation between the behaviourist proposals of B.F. Skinner (1938) and Pavlov (1927), and the cognitive approach proposed by Piaget (1936). While the first theory states that the individual learns fundamentally through direct (be it positive or negative) reinforcement and conditioning, the latter postulates the importance of the individual's mental processes (memorisation, analysis, attention) in learning.

That being said, Bandura's SL does not postulate that behaviourism or cognitivism is to be disregarded, but rather builds upon these theories. Social learning can be understood as a process started by an identificatory event (observation of a model behaviour leading to its emulation) which is memorised, processed, and assigned meaning. That leads to full or partial imitation of such behaviour, which is met with external feedback and observation of the effect produced. Positive feedback or satisfactory outcomes of the emulated behaviour stimulate the individual to further reproduce the model.

4. Learning by Modelling

Social Learning through Model observation is composed of four distinct sequential stages: Attention, Retention, Reproduction, and Motivation *(Bandura, Self-efficacy: Toward a unifying theory of behavioral change.*, 1977). (Delete 'Self-.... change; Bandura, 1977 to be unitalised)

Attention: To learn through observation, the individual must ascertain a determined level of interest in the action/subject observed. This is influenced by a variety of factors, but it can be generalised with how the observer identifies and relates with the observed.

Retention: In order for the behaviour to be set up for reproduction, the observant must retain it in memory, through either visual and/or verbal coding. By observing a repetitive behaviour, they create a mental schema of it, yet it is through verbal encoding that the best results are achieved. When following a recipe, observers are more likely to memorise each step if they have been verbalised rather than just by mere observation (Hacker, 1998).

Reproduction: Behavioural reproduction can only take place if the appropriate conditions have been met (see above) and if the individual is capable of executing them. This can lead to the necessity of reiteration, as some observed behaviours rely on motricity, and the individual will most likely not be able to emulate them on the first try. The process of model observation to reproduction often requires some experimentation, before one is comfortable to implement it and able fully emulate it.

Motivation: The final stage of behaviour appropriation is perhaps the most significant, as it defines if the observer will adopt the behaviour. This is influenced by a variety of factors, such as ideology, cultural background, and the individual's social environment. Above all, a behaviour will be adopted from a model if they "find (it) self-satisfying and (they) reject what they personally disapprove" (Bandura, 1976, citing Hicks, 1971).

Having understood the basic concept of behaviourism and social learning, we can observe their occurrence in everyday life, as well as their effects and possible applications.

5. Citizen Engagement in Future Cities

In 2022, humanity has reached a milestone of eight billion people of population. According to the prognosis, it will reach 9 billion by 2040 and peak numbers of 10 billion in 2060 (Gapminder, Globalis, Hyde, Gapminder, and Gapminder, 2017). To meet the needs of the constantly developing society, the United Nations published a list of Sustainable Development Goals (SGDs), one of them being as follows: "to make cities inclusive, safe, resilient and sustainable" (Nations, 2023). It can be understood as manifesting a need of creating a larger number of smart cities, which would use the most efficient, technologically advanced, and development-oriented resources in order to reach such a goal.

6. How is Social Learning Understood in the Context of Citizen Engagement in Future Cities?

A future city builds upon the concept of a smart city, by encompassing not only the employment of electronic devices and data management to further increase interactivity and productivity, the investment in infrastructure to ensure balanced development or enforcing the sustainability of ecological systems, but it also provides the means to raise public awareness and gather communal intelligence and input through citizen engagement (Azkuna, 2012). To achieve this, an understanding is required of how citizen engagement influences the management of smart/future cities.

There has been substantial amount of research since the introduction of the social learning theory, leaving a need to adapt the definition to the contemporary environment. Social learning is a theory originating in the 1970s, whereas the concept of a smart city was first introduced in the 1990s and has been broadly discussed since. Considering how quickly technology has evolved in the last 50 years, the environmental impact of human activity since the industrial revolution and the constant societal changes, the definition for social learning also needs to be adapted, not only to the current state of the world, but to the context of smart and future cities.

Adapted to citizen engagement, social learning can be understood as a deliberate attempt to adjust the goals or techniques of policy in response to past experience and new information. In this context, an emphasis is put on learning within the process as a result of dialogue and mutual exchange of information between authorities and citizens. Such interaction is meant to "induce reflection upon preferences in a noncoercive fashion" (Dryzek, 2002), while putting emphasis on learning that stems from the collective knowledge of community.

Practically, social learning in citizen engagement comprises three crucial ideas. Firstly, all stakeholders (here: authorities and citizens) shall be fully included in the management of natural resources, making sure that none of the parties owns all the funds, competencies, or information. For that, mutual collaboration is crucial. Secondly, in order to effectively manage the resource there needs to be a structured organisation established with a goal of long-term participation. Thirdly, as according to Holling (1978) natural resource management is a learning process, creation of such learning space and opportunities must be ensured. For such a space to exist, new behaviours and skills must be learned in order to adapt to change and deal with problems constructively, despite uncertainties or differences.

7. Technology as a Social Learning and Citizen Engagement Facilitator

The aim of new technology is increasing the levels of prosperity, survival, and comfort of human civilization (Lee and Trimi, 2018). From the humble origins of humankind, with the invention of the arrowhead to the overwhelming power and possibilities of Generative Artificial Intelligence, its purpose is to provide ways to facilitate human life.

As communities grow, new technological advancements are required to sustain and regulate them, to provide accessibility of engagement and allow inclusivity and active participation of their citizens (Bannon and Griffin, 2001). "The future of social interaction stands at the intersection of technology, human behaviour, and societal norms, shaping the way individuals connect, communicate, and collaborate." (Basir, 2021). The future of social interaction stands at the intersection of technology, human behaviour, and societal norms, shaping the way individuals connect, communicate, and collaborate.

The following are three examples of technologies and their possible applications presented as practical scenarios in social learning and citizen engagement: Extended Reality and the Metaverse, Artificial Intelligence, and Game-based Engagement. Their choice for this article was based on their past and current popularity (Extended Reality and AI) and the constant presence in the human mindset (Game-based Engagement).

8. Extended Reality and the Metaverse

Extended Reality is a term used to encompass technologies that alter, enhance, or replace the individual's view of the world. These are Virtual Reality (VR), Augmented Reality (AR), and Mixed Reality (MR) (Rauschnabel, Felix, Hinsch, Shahab and Alt, 2022). Each of these acts in a particular way, with VR focussing on the complete replacement of the user's vision with a new dimension and reality, AR allowing the user to access more content or information with a mobile or wearable device, by pointing the lens at a triggering point and MR being somewhere in between them, blending the real world with virtual elements. The Metaverse is a virtual space where participants can engage with the usage of AR/VR technology (Mystakidis, 2022, pp. 468-97). Virtual Reality and the Metaverse allow all stakeholders to take part in the discussion and policy making process, by transcending physical barriers and limitations, thus allowing for a much broader audience to participate in the discussion (Dutta, 2024). The implementation of virtual immersive and interactive fora using such technologies also provides transparency and inclusivity in the policy-making process (Comission, An EU initiative on virtual worlds: a head start in the next technological transition, 2023). (Delete 'An…transition') Public projects can be designed with the participation of the citizens that will use it, allowing them to be part of the ideation process, providing their representatives with dataset on necessity. Consequently, the capabilities of the aforementioned technologies facilitate socially positive learning opportunities. By being placed in the midst of discussion, being provided valuable and applicable data and, in case of Extended Reality, being able to engage with the content of the discussion in point-of-view, citizens can make more informed analysis and decisions about the outcomes of policy and projects. Whether in educational scenarios or in training, Extended Reality allows the observer to participate, observe and emulate behaviours as if they were *in loco* (Kaushik, 2020).

Practical scenario: When discussing the usage of nuclear energy and how it benefits the city's energy infrastructure, citizens can participate in virtual reality tours of a model of the project designed in 3D rendering software, where the project planner and other stakeholders are virtually present to provide information and answer questions. Citizens can become better informed about the safety precautions, the increase in jobs and the lowering of energy prices that the power plant will provide. Alternatively, this interactive virtual tour can be done by modelling an existing power plant in a neighbouring city.

The engagement in future cities can be complemented furthermore with the usage of Augmented Reality and Mixed Reality. Through the use of wearables or smart devices, citizens can interact with city services (Raj, 2023) and structures and obtain relevant information about their surroundings. These can take the form of location-based information feeds, such as manually (smart device) or automatically (wearable) scanning a service to determine current waiting times or acquiring information about said service. This would further extend their current common application of providing additional data, such as with the case of musea, which make use of AR for public engagement (Serravalle, Ferraris, Vrontis, Thrassou, and Christofi, 2019).

Nonetheless, the extent of how AR can be used for citizen engagement is far from one-way retrieval of data. Through AR and geospatial information, citizens can have an active role in city planning by participating in the design process of its structure and layout (Saßmannshausen, et al., 2021). By opening the 'virtual' space for citizens to interact with the city in such ways, they are empowered and become engaged stakeholders of their own wellbeing. The capacity for super-imposition of virtual models on public infrastructure allows the user to understand the impact of change and allows them to communicate ways of improving the community.

Practical Scenario: When developing an area for a public park, an AR/MR application can superimpose tracks to reduce crowding, water spots, benches, defibrillator, and first aid spots, lit and unlit areas, and other facilities. The user can submit a placement proposal and observe and upvote others. This crowdsourcing of ideas leads to a better communication of needs and interests of citizens and allows them to learn by modelling (observing others and more practical/feasible solutions), further engaging with their city's prosperity.

The simplicity of looking or pointing a device at a trigger, combined with the reiterative nature of Extended Reality experiences, are some of the key aspects as to why the AR/MR are useful social training tools, as the user can observe the model and engage with it repeatedly (Koutitas, Smith, and Lawrence, 2020). This leads to better decision-making, since it allows for better analysis and reflection.

9. Artificial Intelligence (AI)

Over the last two years, Artificial Intelligence (AI) has been one of the most ubiquitous terms in tech. As of the time of writing this chapter, the term 'AI' reached its highest popularity index on Google Trends (see chart below), which is not surprising, as Generative AI reached a fervour peak with the announcement of such tools as ChatGPT 5, Gemini, and Sora.

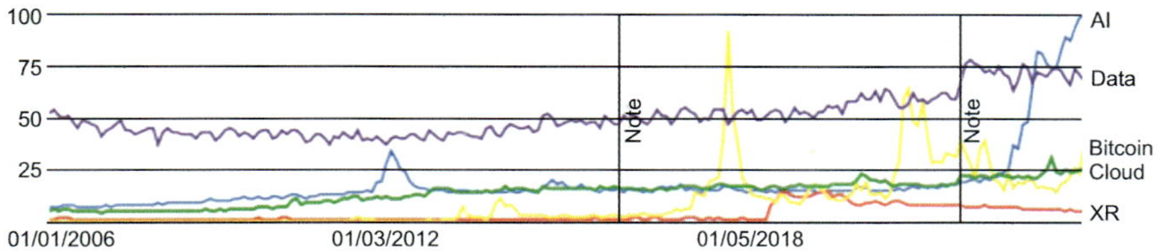

Fig. 1 Interest over time (via Google Searches) from 2006 –2024.
Note: This graph represents the search interest in following terms: AI, XR, Bitcoin, Cloud, and Data. The numbers represent the search interest relative to the highest point on the graph for the specified region and time range. A value of 100 is the peak popularity of the term. A value of 50 means that the term has had half the popularity. A score of 0 means that there was not enough data for this term. Google Trends (n.d.). "AI, XR, Bitcoin, Cloud, Data". Retrieved March 3, 2024, from https://trends.google.com/trends/explore?date=2006-01-01%202024-03-25&q=AI,XR,Bitcoin,Cloud,Data&hl=pt-PT.

An obvious observation is that AI has come to stay. It is now engrained in almost every subject of professional and personal life, either by allowing people to generate content, or by working in the background and allowing daily life to continue to run its course. It is, by definition, a series of processes, logical and mathematical, that aim to mimic the nature of the human brain, allowing connections to be formed between content, subject matter, and data (Chowdhary, 2020).

With the adoption of AI agents and avatars on the rise, which interact with the public in an increasingly seamless and pseudo-human way, they can provide a valued contribution to social learning and citizen engagement (Kyrlitsias and Michael-Grigoriou, 2022). While there is an ongoing discussion about how AI agents are replacing customer service (Huang and Rust, 2018), their application to gauge and acquire data on citizen engagement allows for the community's voice to become more present. This can be achieved with a two-way system. First, an AI avatar can engage with the citizen in active conversation about current community projects and policies and, secondly, the user can initiate topics of discussion with the agent, providing valuable (and anonymous) data to governance. These human-like avatars, running on top of Language Learning Model (LLM) engines, can interact with citizens either via their own personal devices, or in public settings, with accordingly privacy measure.

Practical Scenario: An AI avatar application (displayed, for example, on an LCD screen in a public place) can be interacted with and provide information on upcoming policies, but it can also be used to inquire citizens about their views on them. For example, an Avatar can detect, through computer vision, the presence of a citizen and ask them about the layout of the building they are on, if it is accessible, if there are any modifications to the structure that would benefit their experience.

In the last year alone, significant progress has been made in the field of AI and its capabilities for learning. Training models in AI is expensive and time-consuming (Zhou, Chai, Li and Sun, 2022), but an AI agent capable of determining that an expert behaviour has been observed, in a single experiment, would massively reduce the necessity for its training and, although it is yet to be proven on large-scale social learning scenarios, progress is already being made in the field of AI social learning (Bhoopchand, et al., 2023). The capacity for AI agents to learn, as humans do, from observation and memorization, with task-value attribution and motivation (Bandura, *Social learning theory*, 1977), (Delete 'Social...Theory) opens the possibility of humans learning also from AI agents. It can be postulated that a human, in the nearby future, might observe an AI-powered avatar engage in a specific way and find value and motivation to replicate ecologically friendly or development-oriented behaviours.

The fact that these AI powered personas can take human shapes allows for a stronger engagement (Kyrlitsias and Michael-Grigoriou, 2022). It's been observed that humans, when interacting with robots repeatedly over time, can develop trust in the machine (Kislev, 2022). This human ability has yet to be fully explored, but as the uncanny valley effect becomes less evident, future cities will most likely be filled with AI avatars ready to engage with citizens, allowing them to contribute to a pool of intelligence that capacitates administrators.

10. Games and Game-based Engagement

The interest of humanity in developing and playing games has been long present, way before any individual thought of the concept of digital. It's engrained in the human psyche from an early age to engage in games, either through invitation from others or developed by their own imagination. This propensity for Ludo is one that has been explored already in the many fields, including education (Cheng, Chen, Chu, and Chen, 2015) and marketing, with and without the digital component to great effect (Herbig, 1991, pp. 285-98), as citizens are propelled to engage further with content if done so in a leisurely and organic way, rather than an imposition of ideas (Glas, et al., 2019). As such, for the purpose of this article, Game-based Engagement can be described as the process where, through the usage of digital games, citizens engage in policy-discussion and public project planning, and they interact with their communities for the betterment of themselves and their peers. Digital games can be easily made into social learning tools, where positive and socially beneficial behaviours are encouraged via model observation, peer engagement, and open communication.

Games have definitely crossed the age boundary previously associated with them, being now marketed to all age groups and covering a broad range of interests. In 2022, in Europe, 73% of polled citizens aged 15–24 played digital games and 60% of the 25–34 age group did so as well (Europe, 2022). This data has been underlined to showcase the impact that Game-based Engagement can have on teenagers and young adults, who are more likely to become elements of change in their society (Comission, PROMoting youth Involvement and Social Engagement: Opportunities and challenges for 'conflicted' young people across Europe, 2019). With digital games being a prevalent medium in everyday life of future citizens (Europe, 2022), they should be welcomed as platforms for citizen engagement and prosocial behaviours.

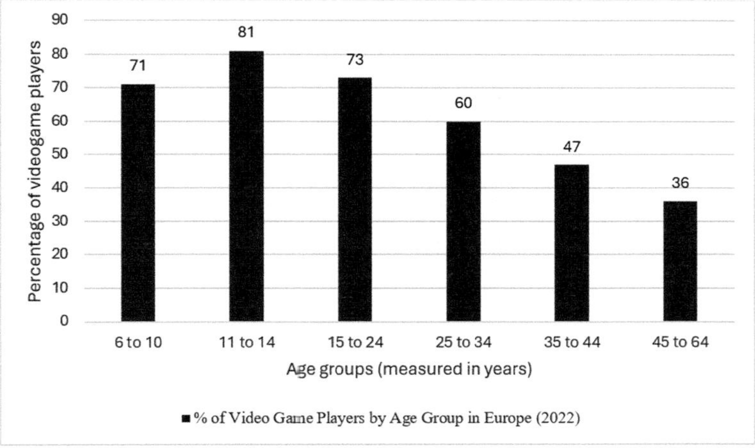

Fig. 2 Distribution of video game players across age groups in years.
Note: Adapted from Europe, V.G. (2022). All About Video Games – European Key Facts. Tech. rep., Video Games Europe.

The processes through which games can be used to achieve citizen learning and engagement vary. When engaging with a game, the individual expects their effort to be rewarded. This can be a simple stimulus from a task, such as pressing a button to execute an action or more complex scenarios, such as developing a functional road and public transport system that reduces traffic in city centres. In particular, games that allow for multiplayer or shared experiences provide environments that potentiate discussion and engagement, and sandbox-based games provide space for visualisation, creation and brainstorming.

As of 2022, the average player in Europe is 32 years of age (Europe, 2022), meaning that gaming culture (casual, hobby, or competitive) has been a part of their development and has become common language. Future cities can make use of it as a communication tool by harnessing the power of games' interactivity to gather and share information, promote socially positive behaviours, and engage further with their citizens.

As such, Game-based Engagement can be used to both crowdsource ideas and to provide citizens with learning opportunities. With software, previously developed or tailor-made for a specific necessity, administrators can interact with the community to share potential zoning projects. Users can provide suggestions using the interactive nature of the game, but also further understand how certain elements will affect the future of their city, for instance, how much the carbon footprint will be increased or decreased. One such example can be observed in Figure 3, which features an interactive game from Finland's Heureka Science Centre "Electri City" exhibit. The game in question shows how policy can affect the environment, energy prices and how society perceives the changes made, since they affect it directly.

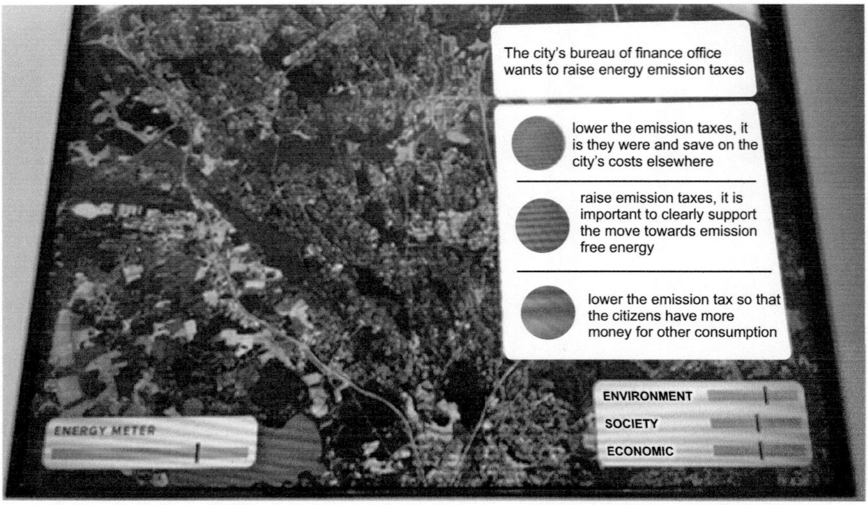

Fig. 3 City Energy Management Game - Heureka Science Centre.
Note: Photograph from an interactive game present at Finland's Heureka Science Centre "Electri City". The user can interact with the elements on screen, taking decisions on policy, and observing their effects in real-time. Photograph taken by the author on March 13, 2019.

The interactive component and fast response of games allow citizens to quickly obtain information on the effects of policy and projects, as games are designed to provide real-time feedback. By using games that provide citizens with the opportunity to share their in-game projects with others, future cities encourage them to learn together and communicate over different perspectives, subsequently empowering them to make better unified policy-making decisions.

Practical Scenario: Using a city-building game, administrators can provide a representation of the city and instructions on how to add a new subway line. The game shows the cost of the line, the time it will take to build, and the concessions required of citizens in terms of daily commute. It can show that this project will affect in-game citizen morale, as people will have their daily routine altered and more traffic jams during the build period. That being said, users can also foresee a decrease in the carbon emissions, as the new line will remove the need to use personal transportation for the daily commute and encourage the usage of cleaner energy sources instead. It will also reduce the daily expenses of citizens as public transportation is overall less costly. By allowing players to alter certain criteria over the construction and share their findings with others, a better understanding of the project can be found. By crowdsourcing knowledge this way, administrators provide an organic learning procedure, where citizens are engaged and see value in their contributions.

Conclusion – The Significance of Social Learning and Citizen Engagement in Future Cities

As society becomes more engaged with their own wellbeing, future cities' administrators will have to recognize the value of crowdsourcing, further than they have before, by using different technologies and approaches. As Balasubramaniam (2017) states in *I, the citizen: Unravelling the power of citizen engagement*:

The government needs to understand that groups or communities that sustain over a period of time are often the ones that evolve organically. The corollary is that expecting community action to sustain through the imposition of orders with uniform guidelines from the top are bound to meet with failure. (...) Community engagement is a subtle and long drawn-out process, which cannot be enforced externally. Nor can one expect communities to respond or perform in a pre-determined manner. One can, at best, work towards creating a conducive environment for communities to dialogue. **(p. 97)**

The usage of modern technology to facilitate resident participation and the examples provided shine a light on how Social Learning and Citizen Engagement can manifest in future-city scenarios. They illustrate how a connected society, where the citizen is given access to participate in the policy decision-making process using different and more accessible means, provides a sense of community, where solutions are crowdsourced for the benefit of all. It is clear that the structure of future cities will be technology; however, their foundations are undoubtedly the voices of their citizens, as these are the ones that will shape the world of tomorrow.

References

Azkuna, I. (2012). *Smart Cities Study: International Study on the Situation of ICT, Innovation and Knowledge in Cities*. The Committee of Digital and Knowledge-based Cities of UCLG. Bilbao.

Baeumler, A., D, Aoust, O., Das, Maitreyi. Bordia, . . . and Terraza, H. (2021). Demographic Trends and Urbanization. *In*: Luc Christiaensen and Nancy Lozano-Garcia (Eds.), *Migrants, Markets, and Mayors: The Employment Challenge in Africa's Secondary Cities*. Washington, DC: World Bank. doi:10.1596/978-1-4648-1112-9. Bandura, A. (1977). Self-efficacy: Toward a Unifying Theory of Behavioral Change. *Psychological Review*, *84*, 191–215. doi:10.1037/0033-295x.84.2.191.

Bandura, A. (1977). *Social Learning Theory*. Englewood Cliffs, New Jersey: Prentice-Hall.

Bannon, L.J. and Griffin, J. (2001, February). New technology, communities, and networking: Problems and prospects for orchestrating change. *Telematics and Informatics*, *18*, 35–49. doi:10.1016/s0736-5853(00)00021-6.

Basir, F. (2021). The Algorithmic Gaze: Surveillance, Privacy, and the Future of Social Interaction. *Research Journal of Psychology*, *1*, 98–104.

Bhoopchand, A., Brownfield, B., Collister, A., Dal Lago, A., Edwards, A., Everett, R., . . . and Zhang, L.M. (2023, November). Learning few-shot imitation as cultural transmission. *Nature Communications*, *14*. doi:10.1038/s41467-023-42875-2.

Cheng, M.-T., Chen, J.-H., Chu, S.-J. and Chen, S.-Y. (2015, July). The use of serious games in science education: A review of selected empirical research from 2002 to 2013. *Journal of Computers in Education*, *2*, 353–75. doi:10.1007/s40692-015-0039-9.

Chowdhary, K.R. (2020). *Fundamentals of Artificial Intelligence*. Springer India. doi:10.1007/978-81-322-3972-7.

Commission, E. (2019). PROMoting youth Involvement and Social Engagement: Opportunities and challenges for 'conflicted' young people across Europe. *PROMoting Youth Involvement and Social Engagement: Opportunities and Challenges for 'Conflicted' Young People across Europe*. Retrieved from https://cordis.europa.eu/project/id/693221.

Commission, E. (2023, July). *An EU Initiative on Virtual Worlds: A Head Start in the Next Technological Transition*. Retrieved from https://digital-strategy.ec.europa.eu/en/library/eu-initiative-virtual-worlds-head-start-next-technological-transition.

Dutta, S. (2024). *Immersive Realm of Extended Reality* (1st Edn.). Delhi: BPB Publications.

Europe, V.G. (2022). *All About Video Games–European Key Facts*. Tech. rep., Video Games Europe.

Faggin, F., Hoff, M.E., Mazor, S. and Shima, M. (1996). The history of the 4004. *IEEE Micro, 16*, 10–20. doi:10.1109/40.546561.

Gapminder, G.S., Globalis, Hyde, Gapminder and Gapminder. (2017, April 6). United Nations –World Population Prospects (2022) –with major processing by Our World in Data. *Systema Globalis*.

Glas, R., Lammes, S., Lange, M., Raessens, J., Vries, I., Jansz, J., . . . and Vries, I. (2019). *The Playful Citizen: Civic Engagement in a Mediatized Culture*. Amsterdam University Press. doi:10.5117/9789462984523.

Hacker, D.J. (Ed.). (1998). *Metacognition in Educational Theory and Practice*. Mahwah, NJ [u.a.]: Erlbaum.

Herbig, P.A. (1991, January). Game theory in marketing: Applications, uses and limits. *Journal of Marketing Management, 7*, 285–98. doi:10.1080/0267257x.1991.9964157.

Huang, M.-H. and Rust, R.T. (2018, February). Artificial Intelligence in Service. *Journal of Service Research, 21*, 155–72. doi:10.1177/1094670517752459.

Kaushik, A. (2020). *XR for Social Impact: A Landscape Review*. Games for Change, NewYork State Council on Arts, NY.

Kislev, E. (2022). *Relationships 5.0*. New York, NY: Oxford University Press.

Kornbluh, A.T. (2003, May). Cities of the Future/the Future of Cities. *Journal of Urban History, 29*, 483–93. doi:10.1177/0096144203029004007.

Koutitas, G., Smith, S. and Lawrence, G. (2020, April). Performance evaluation of AR/VR training technologies for EMS first responders. *Virtual Reality, 25*, 83–94. doi:10.1007/s10055-020-00436-8.

Kyrlitsias, C. and Michael-Grigoriou, D. (2022, January). Social Interaction with Agents and Avatars in Immersive Virtual Environments: A Survey. *Frontiers in Virtual Reality, 2*. doi:10.3389/frvir.2021.786665.

Balasubramaniam, R. (2017). I, the Citizen: Unraveling the Power of Citizen Engagement, 97. doi:10.7591/9781501712470

Lee, S.M. and Trimi, S. (2018, January). Innovation for creating a smart future. *Journal of Innovation & Knowledge, 3*, 1–8. doi:0.1016/j.jik.2016.11.001.

Mystakidis, S. (2022, February). Metaverse. *Encyclopedia, 2*, 486–97. doi:10.3390/encyclopedia2010031.

Nations, U. (2023). *The Sustainable Development Goals Report 2023*. Retrieved from https://unstats.un.org/sdgs/report/2023/The-Sustainable-Development-Goals-Report-2023.pdf.

Pavlov, I.P. (1927). *Conditioned Reflexes: An Investigation of the Physiological Activity of the Cerebral Cortex*. Oxford University Press.

Piaget, J. (1953). *The Origin of Intelligence in the Child*. Routledge and Kegan Paul Ltd.

Raj, S.B. (2023, June). Monitoring the Social Distance in Smart Cities using Wearable Devices. *UGC Care Journal, 47*(2), No. 6S.

Rauschnabel, P.A., Felix, R., Hinsch, C., Shahab, H. and Alt, F. (2022, August). What is XR? Towards a Framework for Augmented and Virtual Reality. *Computers in Human Behavior, 133*, 107289. doi:10.1016/j.chb.2022.107289.

Saßmannshausen, S.M., Radtke, J., Bohn, N., Hussein, H., Randall, D. and Pipek, V. (2021, June). Citizen-Centered Design in Urban Planning: How Augmented Reality can be used in Citizen Participation Processes. *In*: *Designing Interactive Systems Conference 2021*. ACM, New York. doi:10.1145/3461778.3462130.

Serravalle, F., Ferraris, A., Vrontis, D., Thrassou, A. and Christofi, M. (2019, October). Augmented reality in the tourism industry: A multi-stakeholder analysis of museums. *Tourism Management Perspectives*, *32*, 100549. doi:0.1016/j.tmp.2019.07.002.

Skinner, B.F. (1938). *The Behavior of Organisms: An Experimental Analysis*. Appleton-Century. New Your, USA.

Zhou, X., Chai, C., Li, G. and Sun, J. (2022, March). Database Meets Artificial Intelligence: A Survey. *IEEE Transactions on Knowledge and Data Engineering*, *34*, 1096–1116. doi:10.1109/tkde.2020.2994641.

3 Public Policies for Green Cities in the Digital Age

Maria José Sousa[1*]

1. Introduction

Public policies play a pivotal role in determining the sustainability of green cities. Thompson (2002) defines the concept as an urban environment that enhances public health, improves the quality of life, and fosters new attitudes towards nature and sustainability. Urban green spaces are a vital aspect of this intricate environment and consist of parks, woods, farmlands, roadside trees, and structured green parks, each serving ecological, social, and economic purposes (Bradley, 1995; Shafer, 1999; Tyrvainen, 2001; Lutz and Bastian, 2002; Flores et al., 1998).

While the percentage of green space coverage is a significant indicator of a city's greenness, it alone does not reflect the quality of the green structure or the range of ecosystem services provided. The green plot ratio indicator, proposed by Ong (2003), offers a useful means to assess greening levels in urban design and supports urban planning strategies for sustainable development (Diamantini and Zanon, 2000).

By implementing green city policies, the complexity of managing urban green spaces is reduced, benefiting city ecosystems (Shafer, 1999; Landelma et al., 2000; Pauleit and Duhme, 2000). As a result, the research questions for this project emerge: (Rq1) Which public policies are most instrumental in fostering green cities, and (Rq2) what are the primary green city technologies?

The main objective is to understand the prominent public policies and how they influence green city strategies while integrating various dimensions such as infrastructure, technology, energy, and forestry. The European Union (EU) has notably promoted green strategies through innovation support programs in recent years, wherein public policies have played a significant role.

The chapter's innovation lies in mapping published works on public policies for green cities and identifying actions taken in different countries. This serves as a valuable resource for other countries to replicate successful actions and formulate green city strategies. The chapter discusses green urban policies from a social perspective, emphasizing stakeholder participation. It also consolidates scattered

[1] Iscte-Instituto Universitário de Lisboa
* Corresponding author: maria.jose.sousa@iscte-iul.pt

policy information, facilitating researchers and policymakers in developing green urban policies focused on best practices and innovative technologies.

In conclusion, the chapter's primary aim is to provide a global overview of public policies that encourage green cities and sustainable urban development through technological advancements. It achieves this through a comprehensive literature review, policy analysis, and proposing recommendations for policymakers.

2. Methodology

In this study, the b-On scientific engine was employed to conduct a systematic review of the literature, using the search terms "urban green*", 'technology*', and 'polic*'. The Boolean operator 'and' was used to explore policies and technologies applicable to green cities. To ensure relevance and reduce irrelevant results, the search was focused on titles rather than broad topic fields, as supported by previous research by Alvarez-Garca et al. (2018), Amin et al. (2019), Coetzer et al. (2017), Huang et al. (2020), Leong et al. (2020), and Peng et al. (2018).

The review identified 54 research articles covering the period from 2015 to June 2021. Using VOSviewer software, network analysis was conducted based on this data, and a model with four dimensions was applied to analyze the content of official governmental documents on policies and technologies used in green cities. The four dimensions of the model are as follows:

1. Stakeholders' identification of the problem.
2. Mapping the scope of green city and technology policy.
3. Implementation of the policies.
4. Assessment of the impact on stakeholders.

The validity of the model was assessed through a Delphi study, and details regarding the data analysis procedures are provided in the subsequent sections of the report.

Step 1: Keyword co-occurrence analysis

The verification of themes in this study involved utilizing VOSviewer software to conduct network analysis of keyword occurrences and a Scientometrics study to map knowledge domains. The mapping of knowledge domains through VOSviewer, which involves images, allows for the visualization of structural relationships and growth of scientific knowledge. This visualization reveals intricate links between information units in knowledge clusters, encompassing networks, structures, interactions, intersections, evolution, and derivatives.

Academic publications often rely on keywords as crucial bibliometric indicators because they convey the thematic elements of the work. To analyze and identify the knowledge structure, network and cluster analysis was performed on these keywords, considering the number of times a pair of keywords is mentioned in the papers.

Step 2: Public policy analysis

An inductive content analysis approach (Bardin, 1977) was employed to examine various aspects related to the implementation of green technologies in urban afforestation problems. This analysis encompassed the role of governments in policy implementation, stakeholders involved, social participation in green urban issues, and the use of public policies to understand values and interests and support political decisions.

To carry out this analysis, the operational procedure involved applying Berg's method of content analysis (1989), which systematically and objectively identifies special characteristics of messages found in policy reports. The identified traits were then classified and examined as aspects of public policies for green cities using the MAXQDA-designated program.

The analysis followed specific criteria defined by Bardin (1977) for the categories/dimensions, including: (a) Validity, pertinence, or appropriateness; (b) Inclusiveness or completeness; (c) Homogeneity.

Structuring in a single dimension of analysis, ensuring mutual exclusivity for each element to belong to only one category. Objectivity, consistency, or reliability.

After the categorization and examination of the categories, the policy analysis approach proposed by El-Jardali et al. (2014) and Eden et al. (1977) was employed to identify relevant recommendations for policymakers. Notably, the analysis relied on publicly available official papers from EC and National bodies to clarify the origins of the policies under review.

3. Systematic Literature Review

The procedures employed in this systematic review were based on the PRISMA approach, which is an evidence-based framework for reporting in systematic reviews and meta-analyses. The PRISMA methodology encompassed various aspects, including the background of the investigation, the main objectives, the data sources and eligibility criteria, the methodologies used, the results obtained, and finally, the limitations, conclusions, and implications drawn from the findings.

3.1 Eligibility Criteria:Inclusion and Exclusion Criteria

In March 2018, researchers utilized b-on, a scientific information research tool, to conduct an extensive search across various online scientific databases. The search inquiries were based on the terms "urban green", 'technolog', and 'polic'. The study employed specific selection criteria, including:

(a) Focusing on studies related to green city technologies.
(b) Imposing language constraints, limiting the search to articles written in English.
(c) Considering only fully accessible articles.
(d) Including articles that were published after the year 2015.

Table 1 Number of chapters found per query.

Keywords "urban green*", and 'technolog*", and 'polic*'
Number of Scientific papers 2,376
Number of Scientific papers in *Science Direct* since 2015, (2015–2021) 172
Peers reviewed journals 99
Language English 87
Feasibility: 54

3.2 Results of the Papers Search

Table 1 illustrates the number of papers retrieved using the specified search terms. Notably, the number of papers significantly decreased from 2,376 to 172 after introducing the time criterion and exclusively utilizing the Science Direct database (Table 1).

The final selection criteria involved considering only peer-reviewed studies written in English, resulting in 99 papers. Subsequently, the study focused on 54 papers ($n = 54$) for the current investigation. The research addressed two primary questions: (rq1) identifying the most crucial public policies fostering green cities, and (rq2) determining the key green city technologies.

The literature review identified the main dimensions of Green Cities policies, along with the associated technologies, which are presented in Table 2. This table also includes the names of the authors of the papers considered in the literature review.

Table 2 Dimensions of Green Cities Policies, and Technologies by authors.

Dimensions	*Technologies*	*Authors*
Innovation	Virtual reality and Augmented Reality	(Affolderbach et al., 2018; 2019) (Aldy, 2016) (Altenburg et al., 2016) (Barbose et al., 2015) (Bergquist et al., 2015) (Birch and Calvert, 2015) (Brudermann and Sangkakool, 2017) (Bryson et al., 2016) (Cavdar and Aydin, 2015) (Chiu, 2017) (Cohen and Bordass, 2015) (Cowley et al., 2018) (de Boer et al., 2018)
	Artificial Intelligence, Robotics, and Drones	
	Green technologies, green energy	
	Digital platforms, green digital products	
Digital technologies	Artificial Intelligence, Internet of Things, Big Data Analytics	
	Intelligent Systems, Digital platforms, green digital products, Big Data Analytics	

Contd.

Table 2 *Contd.*

<table>
<tr><th>Dimensions</th><th>Technologies</th><th>Authors</th></tr>
<tr><td rowspan="3">Environment</td><td>Internet of Things, Blockchain Technology, Smart Cities Management Technologies</td><td rowspan="8">(Dey, 2016)
(Drummond and Ekins, 2016)
(Dvarioniene et al., 2015)
(Dyckman, 2016)
(Ferrara, 2015)
(Geth et al., 2015)
(Giezen et al., 2018)
(Hammond et al., 2017)
(Hannon et al., 2015)
(Hittmar et al., 2015)
(Hodgson et al., 2016)
(Holley, 2016)
(Irga et al., 2017)
(Kanniah and Siong, 2017)
(Kelly-Detwiler, 2015)
(Kim and Han, 2015)
(Kulkarni et al., 2017)
(Lindman et al., 2016)
(Littlechild, 2016)
(Manders et al., 2016)
(Muscio et al., 2015)
(Newbery, 2016)
(Peyravi, 2015)
(Pitkanen et al., 2016)
(Pittens et al., 2015)
(Rainville, 2017)
(Raunbak et al., 2017)
(Roper and Hewitt-Dundas, 2015)
(Rubashkina et al., 2015)
(Ruby, 2015)
(Scarpellini et al., 2016)
(Schmitz and Altenburg, 2016)
(Schweber et al., 2015)
(Seeberger et al., 2016)
(Sgobbi et al., 2016)
(Strachan et al., 2015)
(Verhees et al., 2015)
(Vermunt et al., 2018)
(Wentworth, 2017)
(Wright et al., 2016)
(Yu et al., 2017)</td></tr>
<tr><td>Internet of Things, Water Intelligent Systems Management, Big Data Analytics</td></tr>
<tr><td>Waste Management Systems, Internet of Things, Blockchain</td></tr>
<tr><td rowspan="5">Rural integration</td><td>Internet of Things, Intelligent Systems for Smart Agriculture, Green energy, Drones, and Robotics</td></tr>
<tr><td>Internet of Things, Drones, and Robotics</td></tr>
<tr><td>Internet of Things, Drones, and Robots, Intelligent transports</td></tr>
<tr><td>Artificial Intelligence, Water Intelligent Systems Management, Drones, and Robotics.</td></tr>
<tr><td>Intelligent Technology for heritage monitoring and Management.
Big Data Analytics.</td></tr>
</table>

The policies promoting green cities are built upon four key dimensions: innovation, digital technologies, environmental considerations, and rural integration.

Within these policies, various technologies play a significant role in driving progress towards green cities. Notable among them are virtual reality and augmented reality, artificial intelligence, robotics, and drones. Additionally, green technologies, green energy solutions, digital platforms, and eco-friendly digital products contribute to the transformation.

The integration of the internet of things, intelligent systems, and blockchain technology further enhances sustainability efforts. Smart cities' initiatives, encompassing management technologies, water intelligent systems, and waste management systems, play a vital role in creating sustainable urban areas. Furthermore, intelligent systems applied in smart agriculture, transportation, heritage monitoring, and big data analytics contribute significantly to the advancement of green cities.

4. Data Network and Discussion

The analysis of all the papers was conducted using Mendeley (Elsevier), resulting in a final database of 54 papers. This database was saved in **RIS** format and uploaded to VOSviewer for keyword occurrence analysis (Figure 1). VOSviewer is a specialized program that facilitates the creation of networks and the evaluation of the strength of associations between variables.

As a result, the strongest link strength between the keywords, based on equal distance, is significant.

5. Policy Analysis of Green Cities Being Reshaped by Technology

The evaluation of public policies has emerged as a prominent global concern, particularly in European Union countries, concerning the sustainability of cities. These policies play a crucial role in supporting initiatives that enable countries, municipalities, and organizations to make more effective investments in creating environmentally friendly cities.

The literature review and examination of various sources, including official reports, documents, and discussions, facilitate the identification of pivotal technologies for green cities (Table 3). This identification can be accomplished through various means, such as visual representations like models, graphics, or data visualization tools (a). Additionally, oral methods, such as recording conversations and engaging in face-to-face interactions (b), as well as audio-visual techniques, such as watching videos and participating in videoconferences (c), can provide valuable insights. Moreover, employing a combination of these methods (d) can offer a comprehensive understanding of the subject matter.

Table 3 displays the technologies supporting Green Cities applications, focusing on urban thermal energy networks, energy efficiency investments, and the adoption of renewable sources for self-consumption. These technologies target improved energy performance in buildings and equipment within public

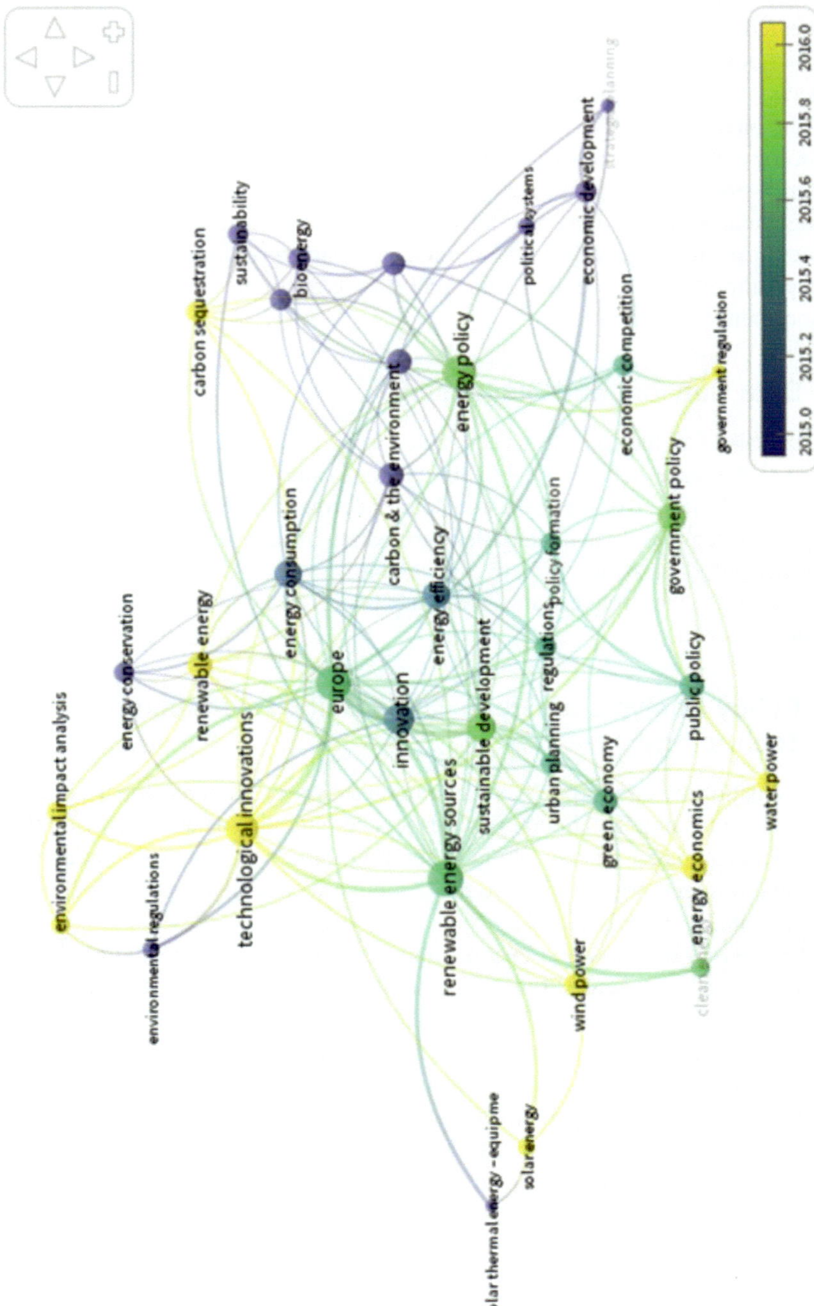

Fig. 1 Keywords occurrence network.

Table 3 Technologies applied to Green cities.

Confrontation of Technologies from the Literature Review and the Policy Analysis

Dimensions	*Public Policy*	*Technologies*
Innovation	Encourage and facilitate the creation of local entrepreneurship and urban innovation networks and hubs, with a particular emphasis on piloting and demonstrating innovative concepts in specific areas. Establish urban living laboratories, business incubators, and innovation centers to foster creativity and experimentation. Additionally, promote the seamless integration of business and technology parks within urban settings, fostering collaboration and advancement in innovative ventures.	Virtual reality and Augmented Reality
	Enhance the connection between cities and their coastal areas, riverine zones, and fishing centers, thereby bolstering the blue economy of coastal urban regions. This multifaceted approach encompasses key aspects such as blue energy, aquaculture, sports, recreational activities, maritime tourism, and cruises, all of which play a prominent role in the sustainable development and economic prosperity of these coastal areas.	Artificial Intelligence, Robotics, and Drones
	Embrace the green economy to implement sustainable development, recognizing the significant impact of public procurement in this pursuit. Encourage companies and institutions within the city to adopt low-carbon strategies, prioritize environmental protection, and promote resource efficiency. Moreover, foster innovation, research, and development of business models, production processes, and environmentally sustainable products.	Green technologies, green energy
	Promote the adoption of innovative urban solutions, both social and technological, specifically tailored for urban spaces. This collaboration with civil society will facilitate the provision of goods and services that align more effectively with the local demand.	Digital platforms, green digital products
Digital technologies	Devise and deploy intelligent systems to monitor and comprehensively manage urban subsystems, driving advancements in fairness, effectiveness, and dependability.	Artificial Intelligence, Internet of Things, Big Data Analytics
	Create electronic solutions to enhance urban governance and optimize urban functions. These solutions encompass information and access platforms for public services, promoting citizen participation and collaboration among various urban stakeholders. Additionally, these solutions will introduce novel employment and commercial models, carefully tailored to meet the distinct needs and capabilities of different sections of the population.	Intelligent Systems, Digital platforms, green digital products, Big Data Analytics

Contd.

Table 3 *Contd.*

Dimensions	*Public Policy*	*Technologies*
Environment	Encouraging the development, enhancement, integration, and accessibility of urban outdoor spaces to create cohesive systems of collective areas. These spaces encompass squares, tree-lined sidewalks, pedestrian zones, traffic-calmed areas, vegetable gardens, parks, and recreational spaces, emphasizing their significance in providing free areas for recreation, leisure, social interaction, cultural activities, and sports. Additionally, recognizing their role in improving microclimates, supporting ecosystems, biogeochemical cycles, biodiversity, and landscapes, thus contributing significantly to green infrastructure.	Internet of Things, Blockchain Technology, Smart cities Management Technologies
	Mitigate urban impact on water resources by prioritizing the enhancement of urban water balance and effective water stress management. This involves optimizing public, domestic, and industrial water consumption, as well as promoting the reuse of graywater, rainwater, and the eco-restoration of industrial effluents.	Internet of Things, Water Intelligent Systems Management, Big Data Analytics
	Enhance urban metabolism efficiency with a primary focus on reducing and repurposing waste as a valuable resource. This involves broadening the spectrum of solutions for waste reuse, recycling, energy and organic recovery, while encouraging the consumption of locally sourced products and combatting wasteful practices.	Waste Management Systems, Internet of Things, Blockchain
Rural integration	Facilitate the integration between urban centers and their rural peripheries, encompassing agricultural and forestry zones along the urban-rural interface. This involves leveraging the economic, social, and cultural synergies that arise from this close proximity. Enhancing transportation and logistics infrastructure will further foster the supply chain for regional production, with a special focus on bolstering the fruit and vegetable sector.	Internet of Things, Intelligent Systems for Smart Agriculture, Green energy, Drones, and Robotics
	To deter the uncontrolled spread of scattered buildings in rural areas, particularly for residential purposes, in order to safeguard urban economies and encourage the optimal utilization of these lands for productive pursuits, such as agriculture and forestry. By discouraging their abandonment and countering opportunistic exploitation, the aim is to support sustainable land management practices.	Internet of Things, Drones, and Robotics
	Improve metropolitan wild spaces by establishing new or revitalizing existing recreational and leisure parks, versatile forest parks, and interconnected route networks in areas surrounding cities. This initiative aims to raise awareness among urban populations about the significance of natural values.	Internet of Things, Drones, and Robots, Intelligent transports

Contd.

Table 3 *Contd.*

Dimensions	*Public Policy*	*Technologies*
	Encourage investments in urban green infrastructure and the interfaces between urban, rural, and natural areas, with a focus on leveraging the benefits of natural capital and the social, economic, and environmental services offered by ecosystems. Foster urban integration by valuing and enhancing river, lagoon, and marine ecosystems, along with associated economic activities.	Artificial Intelligence, Water Intelligent Systems Management, Drones and Robotics.
	Enhance the economic and social appreciation of natural heritage, protected areas, and nature conservation sites by promoting products and services linked to these areas within urban environments. Strengthen their pivotal role in safeguarding biodiversity and asserting the significance of the city-region's natural assets.	Intelligent Technology for heritage monitoring and Management. Big Data Analytics.

spaces. They also contribute to the restoration, maintenance, and development of ecological urban systems and green infrastructure, as well as the certification of public spaces, air quality monitoring, and noise reduction strategies.

Furthermore, these technologies enhance equipment efficiency, such as lighting, windows, insulation, green heating, and energy management systems in commercial buildings. They are aided by artificial intelligence, robotics, and drones to perform tasks challenging for humans, especially related to implementing vertical farming policies. Virtual reality and augmented reality are utilized for creating prototypes and analyzing the best context for public policy implementation.

To combat pollution and enhance citizen quality of life, monitoring the blue ocean economy, water systems, and green cities is essential. Digital platforms and products play a vital role in reducing carbon emissions by cutting traffic, fostering technological development, and promoting innovative work and management models. Blockchain technology, the internet of things, big data analytics, and intelligent systems support improved city management, waste and water management systems, as well as the optimization of energy usage in Green city sites.

Public policies face challenges in green cities technologies adoption for businesses and citizens. State policies focus on intelligent specialization, knowledge transfer, social innovation, and public-interest applications in cities and urban regions. Effective implementation of sustainable urban development policies relies on utilizing available finances. The European Commission prioritizes low-carbon economy promotion, environmental preservation, resource efficiency, risk management, climate change adaptation, competitiveness, innovation, urban regeneration, and social inclusion.

Funding for sustainable urban development initiatives includes the European Social Fund (ESF), Cohesion Fund (CF), European Regional Development Fund (ERDF), and the multi-fund funding theory. The ERDF mandates allocating at least 5% of the fund to sustainable urban development initiatives. A common reference framework is required to verify investments in this area. Operational programs (OP) funded by the European Structural and Investment Funds (ESIF) encompass regional and thematic OPs, such as Competitiveness and Internationalization, Social Inclusion and Employment, Human Capital, and Sustainability and Efficiency in the Use of Resources.

European-level investment solutions like Horizon 2020, INTERREG (Europe, MED, SUDOE, POC-TEP), Atlantic Area, COSME, LIFE, and URBACT III, along with the European Fund for Strategic Investments, complement ESIF financing sources. These efforts aim to enhance the sustainability of cities and urban systems.

Portugal has been actively involved in various projects focusing on sustainable urban development, utilizing funds from multiple sources such as the European Social Fund (ESF), Cohesion Fund (CF), European Regional Development Fund (ERDF), and other European-level investment programs. Here are some relevant projects and initiatives where Portugal is or was a partner:

1. URBACT III

URBACT III is a European exchange and learning program promoting sustainable urban development. Several Portuguese cities participate in URBACT networks, such as:

- **Lisbon** in the "Creative Spirits" network, focusing on integrating the cultural and creative sectors into local development policies.
- **Porto** in the "CityMobilNet" network, which aims to develop sustainable urban mobility plans.

2. Horizon 2020

Portugal is involved in multiple Horizon 2020 projects that contribute to sustainable urban development. Examples include:

- **Sharing Cities:** This project aims to implement innovative digital solutions in urban areas, focusing on reducing carbon emissions and improving energy efficiency.
- **REPLICATE:** Porto is a pilot city in this project, which focuses on smart city technologies to enhance energy efficiency, mobility, and ICT.

3. INTERREG Programs

Portugal participates in various INTERREG programs, which facilitate cross-border cooperation to address shared regional challenges. Key programs include:

- **INTERREG Europe:** Portuguese cities and regions collaborate with European partners on projects that support sustainable urban development.
- **INTERREG Atlantic Area:** This program promotes transnational cooperation along the Atlantic Ocean, with projects focused on innovation, resource efficiency, and sustainable urban planning.

- **INTERREG MED:** Focuses on cooperation in the Mediterranean region, addressing urban challenges such as energy efficiency, waste management, and sustainable mobility.

4. COSME

The COSME program supports the competitiveness of enterprises, with a particular focus on SMEs. Portugal is involved in projects that enhance urban sustainability through innovative business solutions and public-private partnerships.

5. LIFE Program

Portugal participates in the LIFE program, which funds environmental and climate action projects. Examples include:

- **LIFE PAYT:** Aims to develop and implement a pay-as-you-throw waste management system in several municipalities, improving waste collection efficiency, and recycling rates.
- **LIFE SMART Waste:** Focuses on innovative approaches to waste management and circular economy practices.

6. European Fund for Strategic Investments (EFSI)

Portugal benefits from EFSI, which aims to mobilize private investment in strategic projects. Examples include infrastructure projects that enhance urban sustainability, such as smart grid developments and urban mobility improvements.

7. Operational Programs (OPs) under ESIF

Portugal implements several operational programs funded by ESIF, which target sustainable urban development:

- **PO SEUR (Operational Program for Sustainability and Efficient Use of Resources):** Focuses on promoting resource efficiency, urban mobility, and environmental protection.
- **POCI (Operational Program for Competitiveness and Internationalization):** Supports innovation and competitiveness, including urban development projects that integrate smart technologies and sustainable practices.
- **POISE (Operational Program for Social Inclusion and Employment):** Funds projects that enhance social inclusion and employment opportunities in urban areas.
- **PO CH (Operational Program for Human Capital):** Invests in education and skills development, supporting sustainable urban growth through human capital enhancement.

Portugal's commitment to sustainable urban development is evident through its multifaceted approach, leveraging a range of funding sources and addressing various urban challenges. From urban regeneration and housing to environmental sustainability and technological innovation, these projects and programs are transforming Portuguese cities into more livable, inclusive, and sustainable urban environments. By prioritizing community engagement and participatory planning,

Portugal is ensuring that its urban development initiatives are not only effective but also equitable and responsive to the needs of its citizens.

One of the cornerstones of Portugal's sustainable urban development strategy is urban regeneration. Many Portuguese cities, including Lisbon and Porto, have embarked on extensive urban regeneration projects aimed at revitalizing historic neighborhoods and underutilized urban areas. These projects often involve the restoration of historic buildings, the creation of public spaces, and the improvement of urban infrastructure.

For instance, the Mouraria neighborhood in Lisbon has undergone significant transformation through the Programa de Desenvolvimento Comunitário da Mouraria (Mouraria Community Development Program). Funded by both national sources and the European Union, this program has focused on rehabilitating old buildings, enhancing public spaces, and fostering community engagement. The revitalization of Mouraria has not only preserved the historical heritage of the area but has also stimulated local businesses and tourism, contributing to economic sustainability.

Portugal has also prioritized addressing housing shortages and promoting social inclusion through various initiatives. The Programa 1° Direito (First Right Program) is a key initiative in this regard, aimed at providing access to decent housing for all, particularly for vulnerable populations. This program is funded through a combination of national funds and the European Regional Development Fund (ERDF). By focusing on renovating existing buildings and constructing new housing units, the 1° Direito Program addresses the dual challenge of urban housing shortages and the need for sustainable building practices.

Another program is the Bairro Saudável (Healthy Neighborhood) initiative, which aims to improve living conditions in deprived urban areas. This program, supported by national and EU funds, promotes the rehabilitation of housing, the creation of green spaces, and the development of community facilities. By fostering a sense of community and improving living conditions, Bairro Saudável contributes to social cohesion and the overall well-being of residents.

Portugal is committed to sustainable urban development, utilizing a variety of funding sources to tackle urban challenges and enhance the quality of life in its cities. This commitment is evident in numerous projects and programs that span various aspects of urban life, from infrastructure and housing to environmental sustainability and technological innovation. These initiatives are not only transforming the physical landscape of Portuguese cities but are also fostering social inclusion, economic growth, and environmental stewardship.

One of the basis of Portugal's sustainable urban development strategy is urban regeneration. Many Portuguese cities, including Lisbon and Porto, have implemented extensive urban regeneration projects aimed at revitalizing historic neighborhoods and underutilized urban areas. These projects often involve the restoration of historic buildings, the creation of public spaces, and the improvement of urban infrastructure.

For instance, the Mouraria neighborhood in Lisbon has undergone significant transformation through the Programa de Desenvolvimento Comunitário da Mouraria (Mouraria Community Development Program). Funded by both national sources and the European Union, this program has focused on rehabilitating old buildings, enhancing public spaces, and fostering community engagement. The revitalization of Mouraria has not only preserved the historical heritage of the area but has also stimulated local businesses and tourism, contributing to economic sustainability.

Portugal has also prioritized addressing housing shortages and promoting social inclusion through various initiatives. The Programa 1° Direito is a key initiative in this regard, aimed at providing access to decent housing for all, particularly for vulnerable populations. This program is funded through a combination of national funds and the European Regional Development Fund (ERDF). By focusing on renovating existing buildings and constructing new housing units, the 1° Direito Program addresses the dual challenge of urban housing shortages and the need for sustainable building practices.

Another program is the Bairro Saudável (Healthy Neighborhood) initiative, which aims to improve living conditions in deprived urban areas. This program, supported by national and EU funds, promotes the rehabilitation of housing, the creation of green spaces, and the development of community facilities. By fostering a sense of community and improving living conditions, Bairro Saudável contributes to social cohesion and the overall well-being of residents.

Portugal's commitment to environmental sustainability is reflected in its efforts to enhance green spaces and promote sustainable urban mobility. The country has invested significantly in creating and maintaining urban parks, gardens, and green corridors. The Parque das Nações in Lisbon is a prime example of urban redevelopment that integrates extensive green spaces. This former industrial area was transformed into a green area featuring parks, a marina, and various leisure facilities, making it a model of sustainable urban planning.

In addition to green spaces, Portugal has implemented numerous initiatives to promote sustainable urban mobility. The development of extensive bike-sharing programs and the expansion of public transportation networks are critical components of these efforts. Lisbon's Gira bike-sharing system, supported by municipal funds and private partnerships, has made cycling a viable and sustainable mode of transportation in the city. Similarly, the expansion of metro and tram networks in cities like Porto has reduced reliance on private cars, leading to decreased traffic congestion and lower carbon emissions.

Embracing technological innovation is another key aspect of Portugal's approach to sustainable urban development. The Smart Cities initiative, supported by national and EU funds, encourages cities to adopt digital technologies to improve urban management and service delivery. These technologies include smart lighting, intelligent traffic management systems, and data-driven urban planning tools.

The city of Porto, for example, has implemented various smart city projects that leverage technology to enhance urban life. The Porto Living Lab, a collaborative project involving the municipal government, universities, and private companies, focuses on testing and implementing smart solutions for urban challenges. Projects under the Porto Living Lab include smart parking systems, energy-efficient street lighting, and real-time air quality monitoring.

This approach can serve as a model for sustainable urban development that other countries can implement, demonstrating that with strategic investment and community involvement, cities can become thriving, resilient, and sustainable places to live.

6. Final Considerations

The primary goal of public policies advocating sustainable development is to significantly enhance the quality of life for people. To effectively implement Green Cities initiatives, it is crucial to establish practical instruments for implementation and monitoring. The creation and execution of public policies for green cities heavily rely on established programs and funding, with a focus on national and integrated territorial approaches.

This chapter provides an opportunity to systematize existing research on public policies for green cities and to outline the main policy axes at the European level. These achievements are essential for advancing knowledge as they foster networks and knowledge-sharing platforms, promoting the dissemination of best practices.

Key findings reveal that green cities can bring numerous benefits not only to urban centers but also to the non-urban environment within their functional influence, such as agricultural hinterland areas, surrounding forest spaces, and rural interfaces. Green city policies aim to foster cross-border collaboration by encouraging complementarities, adaptability, and access to major international markets and territories. Additionally, they support the enhancement and integration of urban physical infrastructure, including parks, environmental conditions, and landscapes. Furthermore, these policies contribute to the functional, cultural, social, and economic development of urban areas by seeking compromise solutions between various territorial agents, including public, private, and associative entities.

A key focus of these policies is on enhancing the base of endogenous resources, improving the efficiency of subsystems like energy, mobility, water, and waste management, and bolstering the ability to cope with risks and impacts, particularly those associated with climate change.

References

Affolderbach, J., O'Neill, K. and Preller, B. (2019). Global-local tensions in urban green neighborhoods: A policy mobilities approach to discursive change in Freiburg, Vancouver, and Luxembourg. *Geografiska Annaler: Series B, Human Geography*, 1–20. doi:10.1080/04353684.2019.1681286.

Affolderbach, J., Schulz, C. and Braun, B. (2018). Green building as an urban climate change strategy. *In*: *Urban Book Series*. https://doi.org/10.1007/978-3-319-77709-2_1.

Aldy, J.E. (2016). Mobilizing Political Action on Behalf of Future Generations. *Future of Children*, *26*(1), 157–78.

Altenburg, T., Sagar, A., Schmitz, H. and Xue, L. (2016). Guest editorial: Comparing low-carbon innovation paths in Asia and Europe. *Science and Public Policy (SPP)*, *43*(4), 451–53.

Amin, M.T., Khan, F. and Amyotte, P. (2019). A bibliometric review of process safety and risk analysis. *Process Safety and Environmental Protection*, *126*, 366–81. https://doi.org/10.1016/j.psep.2019.04.015.

Barbose, G., Darghouth, N.R., Weaver, S., Feldman, D., Margolis, R. and Wiser, R. (2015). Tracking US photovoltaic system prices 1998–2012: A rapidly changing market. *Progress in Photovoltaics*, *23*(6), 692.

Bergquist, A.-K. and Söderholm, K. (2015). Transition to greener pulp: Regulation, industry responses, and path dependency. *Business History*, *57*(6), 862.

Birch, K. and Calvert, K. (2015). Rethinking 'Drop-in' Biofuels: On the Political Materialities of Bioenergy. *Science and Technology Studies*, *28*(1), 52–72.

Bradley, G. (1995). *Urban Forestry Landscapes: Integrating Multidisciplinary Perspectives*. University of Washington Press, Seattle.

Brudermann, T. and Sangkakool, T. (2017). Green roofs in temperate climate cities in Europe: An analysis of key decision factors. *Urban Forestry and Green Cities*, *21*. https://doi.org/10.1016/j.ufug.2016.12.008.

Bryson, D., Atwal, G., Chaudhuri, A. and Dave, K. (2016). Antecedents of Intention to Use Green Banking Services in India. *Strategic Change*, *25*(5), 551–67.

Cavdar, S.C. and Aydin, A.D. (2015). An empirical analysis about technological development and innovation indicators. *In*: *World Conference on Technology, Innovation, and Entrepreneurship*, *195*, 1486–95.

Chiu, C.H. (2017). Regulatory regimes and the environmental implications of sustainability-based governance of extra-legal buildings: Comparing the Taichung and Kaohsiung policies. *Journal of Architecture and Planning*, *18*(2)

Coetzer, M., Bussin, M. and Geldenhuys, M. (2017). The Functions of a Servant Leader. *Administrative Sciences*, *7*(1), 5. https://doi.org/10.3390/admsci7010005.

Cohen, R. and Bordass, B. (2015). Mandating transparency about building energy performance in use. *Building Research and Information*, *43*(4), 534–52.

Cowley, R., Joss, S. and Dayot, Y. (2018) The smart city and its publics: Insights from across six UK cities. *Urban Research and Practice*, *11*(1), 53.

de Boer, J., Zuidema, C., van Hoorn, A. and de Roo, G. (2018). The adaptation of Dutch energy policy to emerging area-based energy practices. *Energy Policy*, *117*, 142–50.

Dey, S. (2016). Does a strong patent regime discourage innovation? *Economics of Innovation and New Technology*, *25*(5), 485.

Diamantini, C. and Zanon, B. (2000). Planning the urban sustainable development, the case of the plan for the province of Trento, Italy *Environ. Impact Assess. Rev.*, *20*, 299–310.

Drummond, P. and Ekins, P. (2016). Reducing CO 2 emissions from residential energy use. *Building Research and Information*, *44*(5/6), 585–603.

Dvarioniene, J., Gurauskiene, I., Gecevicius, G., Trummer, D.R., Selada, C., Marques, I. and Cosmi, C. (2015). Stakeholders involvement for energy-conscious communities: The Energy Labs experience in 10 European communities. *Renewable Energy: An International Journal*, *75*, 512–18.

Dyckman, C.S. (2016). Sustaining the Commons: The Coercive to Cooperative, Resilient, and Adaptive Nature of State Comprehensive Water Planning Legislation. *Journal of the American Planning Association*, *82*(4), 327–349.

Ferrara, R. (2015). The Smart City and the Green Economy in Europe: A Critical Approach. *Energies (19961073)*, *8*(6), 4724–34.

Flores, A., Pickett, S.T.A., Zipperer, W.C., Pouyat, R.V. and Pirani, R. (1998). Adopting a modern ecological view of the metropolitan landscape: The case of a greenspace system for the New York City region. *Landscape and Urban Plan.*, *39*, 295–308.

Geth, F., Brijs, T., Kathan, J., Driesen, J. and Belmans, R. (2015). An overview of large-scale stationary electricity storage plants in Europe: Current status and new developments. *Renewable and Sustainable Energy Reviews*, *52*, 1212–27.

Giezen, M., Balikci, S. and Arundel, R. (2018). Using Remote Sensing to Analyse Net Land-Use Change from Conflicting Sustainability Policies: The Case of Amsterdam. *ISPRS International Journal of Geo-Information*, *7*(9). https://doi.org/10.3390/ijgi7090381.

Hammond, G.P. and O'Grady, Á. (2017). The potential environmental consequences of shifts in UK energy policy that impact electricity generation. *Proceedings of the Institution of Mechanical Engineers – Part A – Power & Energy.* Sage Publications Ltd., *231*(6), 535.

Hannon, M.J., Foxon, T.J. and Gale, W.F. (2015). "Demand pull" government policies to support Product-Service System activity: The case of Energy Service Companies (ESCos) in the UK.

http://widgets.ebscohost.com/prod/customerspecific/ns000290/authentication/index.php?url=https%3A%2F%2Fsearch.ebscohost.com%2Flogin.aspx%3Fdirect%3Dtrue%26AuthType%3Dip%2Ccookie%2Cshib%2Cuid%26db%3Da9h%26AN%3D110348265%26amp%3Blang%3Dpt-br%26site%3Deds-li

Hittmar, S., Varmusa, M. and Lendela, V. (2015). Proposal of evaluation system for successful application of innovation strategy through a set of indicators. *Procedia Economics and Finance*, *26*, 17–22. Elsevier.

Hodgson, E., Ruiz-Molina, M., Marazza, D., Pogrebnyakova, E., Burns, C., Higson, A., Rehberger, M., Hiete, M., Gyalai-Korpos, M., Lucia, L.D., Noël, Y., Woods, J. and Gallagher, J. (2016). Horizon scanning the European bio-based economy: A novel approach to the identification of barriers and key policy interventions from stakeholders in multiple sectors and regions. *Biofuels, Bioproducts, and Biorefining*, *10*(5), 508–22.

Holley, C. (2016). Linking Law and New Governance: Examining Gaps, Hybrids, and Integration in Water Policy. *Law and Policy*, *38*(1), 24–53.

Huang, L., Zhou, M., Lv, J. and Chen, K. (2020). Trends in global research in forest carbon sequestration: A bibliometric analysis. *Journal of Cleaner Production*, *252*, 1–17. https://doi.org/https://doi.org/10.1016/j.jclepro.2019.119908.

Irga, P.J., Braun, J.T., Douglas, A.N.J., Pettit, T., Fujiwara, S., Burchett, M.D. and Torpy, F. R. (2017). The distribution of green walls and green roofs throughout Australia: Do policy instruments influence the frequency of projects? *Urban Forestry and Green Cities*, *24*. https://doi.org/10.1016/j.ufug.2017.03.026.

Kanniah, K.D. and Siong, H.C. (2017). Urban forest cover change and sustainability of Malaysian cities. *Chemical Engineering Transactions*, *56*. https://doi.org/10.3303/CET1756113.

Kelly-Detwiler, P. (2015). Power Pushing to the Periphery, Threatening Stability of US Utilities. *Natural Gas and Electricity*, *31*(7), 13–17.

Kim, M.H. and Han, Y. (2015). Investigating the Empirical Relationship between Polity and Economic Growth. *Review of Political Economy*, *27*(3), 341–68.

Kulkarni, G., Nilesh, D., Parag, B., Wasnik, P., Hambarde, K., Tamsekar, P., Kamble, V. and Bahuguna, V. (2017). Effective use of GIS-based spatial pattern technology for urban greenery space planning: A case study for Ganesh Nagar area of Nanded city. *Advances in Intelligent Systems and Computing*, *467*. https://doi.org/10.1007/978-981-10-1645-5_11.

Landelma, R., Salminen, P. and Hokkanen, J., (2000). Using multicriteria methods in environmental planning and management. *Environ. Manage.*, *26* (6), 595–605.

Leong, L.-Y., Hew, T.-S., Tan, G. W.-H., Ooi, K.-B. and Lee, V.-H. (2020). Tourism research progress: A bibliometric analysis of tourism review publications. *Tourism Review*. https://doi.org/10.1108/tr-11-2019-0449.

Lindman, Asa and Söderholm, P. (2016). Wind energy and green economy in Europe: Measuring policy-induced innovation using patent data. *Applied Energy*, *179*, 1351–59.

Littlechild, S. (2016). Contrasting Developments in UK Energy Regulation: Retail Policy and Consumer Engagement. *Economic Affairs*, *36*(2), 118–32.

Lutz, M. and Bastian, O., 2002. Implementation of landscape planning and nature conservation in the agricultural landscape: A case study from Saxony. *Agric. Ecosyst. Environ.*, *92*, 159–70.

Manders, T.N., Höffken, J.I. and van der Vleuten, E.B.A. (2016). Small-scale hydropower in the Netherlands: Problems and strategies of system builders. *Renewable and Sustainable Energy Reviews*, *59*, 1493–1503.

Muscio, A., Reid, A. and Rivera Leon, L. (2015). An empirical test of the regional innovation paradox: Can smart specialization overcome the paradox in Central and Eastern Europe? *Journal of Economic Policy Reform*, *18*(2), 153–71.

Newbery, D.M. (2016). Towards a green energy economy? The EU Energy Union's transition to a low-carbon zero-subsidy electricity system: Lessons from the UK's Electricity Market Reform. *Applied Energy*, *179*, 1321–30.

Ong, B.L. (2003). Green plot ratio: An ecological measure for architecture and urban planning. *Landscape and Urban Plan.*, *63*, 197–211.

Pauleit, S. and Duhme, F. (2000). Assessing the environmental performance of land cover types for urban planning. *Landscape and Urban Plan.*, *52*, 1–20.

Peng, B., Guo, D., Qiao, H., Yang, Q., Zhang, B., Hayat, T., Alsaedi, A. and Ahmad, B. (2018). Bibliometric and visualized analysis of China's coal research 2000–2015. *Journal of Cleaner Production*, *197*, 1177–89. https://doi.org/10.1016/j.jclepro.2018.06.283.

Peyravi, B. (2015). Southeast European Countries: The Needs of Innovations in the Context of Enlargement of The European Union. *Public Administration (16484541)*, *3/4*(47/48), 112–120.

Pittens, C.A.C.M., Vonk Noordegraaf, A., Veen, S.C., Anema, J.R., Huirne, J.A.F. and Broerse, J.E.W. (2015). The involvement of gynecological patients in the development of a clinical guideline for resumption of (work) activities in the Netherlands. *Health Expectations*, *18*(5), 1397–1412.

Rainville, A. (2017). Standards in green public procurement: A framework to enhance innovation. *Journal of Cleaner Production*, *167*, 1029–37.

Raunbak, M., Zeyer, T., Zhu, K. and Greiner, M. (2017). Principal Mismatch Patterns across a Simplified Highly Renewable European Electricity Network. *Energies (19961073)*, *10*(12), 1–13.

Roper, S. and Hewitt-Dundas, N. (2015). Knowledge stocks, knowledge flows, and innovation: Evidence from matched patents and innovation panel data. *Research Policy*, *44*(7), 1327–40.

Rubashkina, Y., Galeotti, M. and Verdolini, E. (2015). Environmental regulation and competitiveness: Empirical evidence on the Porter Hypothesis from European manufacturing sectors. *Energy Policy*, *83*, 288–300.

Ruby, T.M. (2015). Innovation-enabling policy and regime transformation towards increased energy efficiency: The case of the circulator pump industry in Europe. *Journal of Cleaner Production*, *103*, 574–85.

Scarpellini, S., Valero-Gil, J. and Portillo-Tarragona, P. (2016). The "economic–finance interface" for eco-innovation projects. *International Journal of Project Management*, *34*(6), 1012–25.

Schmitz, H. and Altenburg, T. (2016). Innovation paths in Europe and Asia: Divergence or convergence? *Science and Public Policy (SPP)*, *43*(4), 454–63.

Schweber, L., Lees, T. and Torriti, J. (2015). Framing evidence: Policy design for the zero-carbon home. *Building Research and Information*, *43*(4), 420–34.

Seeberger, J., Grandhi, R., Kim, S. S., Mase, W.A., Reponen, T., Ho, S. and Chen, A. (2016). E-Waste Management in the United States and Public Health Implications. *Journal of Environmental Health*, *79*(3), 8–16.

Sgobbi, A., Simões, S.G., Magagna, D. and Nijs, W. (2016). Assessing the impacts of technology improvements on the deployment of marine energy in Europe with an energy system perspective. *Renewable Energy: An International Journal*, *89*, 515–25.

Shafer, C. (1999). US National park buffer zones: Historical, scientific, social, and legal aspects. *Environ. Manage.*, *23* (1), 49–73.

Strachan, P.A., Cowell, R., Ellis, G., Sherry-Brennan, F. and Toke, D. (2015). Promoting Community Renewable Energy in a Corporate Energy World. *Sustainable Development*, *23*(2), 96–109.

Thompson, C. (2002). Urban open space in the 21st century. *Landscape and Urban Plan.*, *60*, 59–72.

Tyrvainen, L. (2001). Economic valuation of urban forest benefits in Finland. *J. Environ. Manage.*, *62*, 75–92.

Verhees, B., Raven, R., Kern, F. and Smith, A. (2015). The role of policy in shielding, nurturing, and enabling offshore wind in The Netherlands (1973–2013). *Renewable and Sustainable Energy Reviews*, *47*, 816–29.

Vermunt, N.P., Harmsen, M., Elwyn, G., Westert, G.P., Burgers, J.S., Olde Rikkert, M.G. and Faber, M.J. (2018). A three-goal model for patients with multi-morbidity: A qualitative approach. *Health Expectations*, *21*(2), 528–38.

Wentworth, J. (2017). Urban Green Infrastructure and Ecosystem Services. *POSTbrief No.26.*, Parliamentary Office of Science and Technology, UK.Wright, N., Rowley, E., Chopra, A., Gregoriou, K. and Waring, J. (2016). From admission to discharge in mental health services: A qualitative analysis of service user involvement. *Health Expectations*, *19*(2), 367–76.

4 Innovative Reporting for Resilient Cities
Bridging Sustainability and Transparency

Silvana Secinaro,[1] Davide Calandra[2] and Ginevra Degregori[3*]

1. Introduction

The literature regarding the public sector presents numerous theories on defining the efficiency of public administration. It is commonly recognised that correctly managing what is commonly known as "*res publica*" through socio-environmental responsibility practices represents an important issue for the State and, consequently, for citizens. It is fundamental for budget management processes to be transparent and understandable, and to consider aspects related to sustainable development in addition to those of mere financial optimisation. The growing awareness of economic, social, and environmental issues has increasingly led public institutions to adopt models in decision-making that focus on economic development and sustainable and inclusive practices. This represents a real social responsibility that is considered part of sustainable development (Střihavková et al., 2021). Furthermore, environmental, social and governance (ESG) aspects have specific implications within the field of public sector risk management (Zioło et al., 2022). In this regard, the world population is growing exponentially year after year, reaching eight billion in November 2022 and simultaneously, an urbanisation phenomenon is occurring where more and more people live in cities. These considerations urge a focus on city management models as it has become necessary to develop new ways and solutions to ensure all citizens' quality of life. In this context, Smart Cities are an integral part of the ideology of society's sustainable development (Artamonov et al., 2020) and, therefore, can represent a possible solution to the previously exposed weaknesses. Indeed, a Smart City represents a kind of urban innovation that creates public value by establishing new ideas and practices (Barrutia et al., 2022). Among the main elements characterising smart cities that can be identified are the transparency and stakeholder engagement issues in the economic, social, and environmental spheres. The central point of Smart Cities concerns constructing a solid system based on the involvement of multiple public and private entities to increase public value (Bolívar, 2018). In this regard, only through the collaboration between the public sector, civil society organisations, and

[1,2,3] University of Turin, Department of Management "Valter Cantino"
Email : silvana.secinaro@unito.it; davide.calandra@unito.it
* Corresponding author: ginevra.degregori@unito.it

the community inclusive innovation will occur successfully (Farneti and Siboni, 2011). To enable such collaboration, the plurality of citizens needs to understand and interpret the information in public administration budgets because only in this way a real government-citizen relationship can be established, leading to an increase in innovation practices and public value. As stated by Montesinos and Brusca (2019), tools that can offer opportunities for improving transparency and accountability of public administrations are citizens' reports, in which financial, social, and environmental aspects are presented clearly and comprehensively. One of these reporting tools can be identified in the *Popular Financial Report* (or simply PFR), which is a document that allows reporting financial and non-financial information in a simple, comprehensible, and accessible way to a wide range of audiences, including people who might not have a specific background in finance or accounting fields. In conjunction with the economic and financial information, the PFR also considers additional data and non-financial information to strengthen the goal of increasing transparency in reporting practices, highlighting the entities' ability to generate economic and social value for the community (Montesinos and Brusca, 2019). PFR can be interpreted as an effective tool able to create a connection between public entities and the community, facilitating the understanding of financial and economic information and, at the same time, clarifying sustainability efforts within the public sector (Biancone et al., 2019). This approach seeks to achieve a dual purpose: on the one hand, it empowers citizens by providing them with accessible information and, on the other, it encourages them to actively participate in the decision-making process, amplifying their voices in public debate (Biancone et al., 2016). In addition, by presenting sustainability information and non-financial data in a more understandable form, PFR can also stimulate public demand for more sustainable urban development practices, leading to prioritise environmental management and equitable resource allocation. All of this can lead to the creation of urban environments that are not only resilient to socio-economic and environmental challenges, but also characterized by greater inclusiveness and fair access to opportunities for all citizens (Brescia, 2019; Grossi et al., 2021).

In conclusion, an innovation that promotes the increase of public value is necessary, considered as community development that, through more effective regulations, better resource allocation, and the creation of programs and initiatives that encourage citizen involvement, can promote sustainable practices. Active citizen involvement is essential to ensure that public choices align with society's needs and values, and to increase public administration's accountability and transparency.

This paper proposes an investigation that includes bibliometric analysis of peer-reviewed articles' data (Dal Mas et al., 2019; Secinaro and Calandra, 2020) collected from the Scopus database. Several interesting observations emerge from this analysis, especially linked to public sector value creation, sustainability efforts, and stakeholder engagement, focusing on Smart Cities and the PFR.

Based on these considerations, this chapter aims to answer three main research questions:

RQ1: What are the sources, keywords, social structure, and conceptual frameworks related to Smart Cities' reporting practices?
RQ2: What are the main theoretical and practical implications of the research?
RQ3: What are the research limitations and possible future directions of the analysed topic?

The paper will be structured as follows. Section 2 provides a review of the existing literature. Section 3 focuses on the methodology adopted to conduct the analysis. Section 4 summarises the main findings that emerged from the analysis. Section 5 discusses the results obtained. Section 6 concludes the article by synthesising the current state of the art and recommendations for future research lines.

2. Literature Review

In recent decades, there has been a growing awareness among the public opinion regarding the importance of sustainable development. This evolution has prompted global financial markets to take these principles increasingly seriously, incorporating them into their valuation and investment models. The concept of 'Sustainability' generally refers to the definition provided by the World Commission on Environment and Development in 1987, that has defined this term as an economic development activity that "meets the needs of the present without compromising the ability of future generations to meet their own needs"(Secretary-General and Development, 1987). Originally with a definition focused purely on ecological themes and aspects, the term 'Sustainability' has expanded to encompass the concept of corporate social responsibility (CSR) and to be analyzed based on qualitative and quantitative evaluations regarding social (S) and governance (G) aspects, in addition to strictly environmental (E) ones. The need to pursue sustainable development has become a central theme (Castellano et al., 2022); this is also due to the increased sensitivity of citizens to the improvement of environmental and social conditions. Following this need, regulatory frameworks have also been identified to provide various organizations guidelines to follow to facilitate their communication of non-financial information, concerning socio-environmental aspects. In 2015, all member states of the United Nations adopted the 2030 Agenda for Sustainable Development, based on the 17 Sustainable Development Goals (SDGs). During the same year, 183 members of the United Nations Framework Convention on Climate Change (UNFCCC) signed the Paris Agreement, an international commitment that aims to preserve global climate stability and maintain a limited increase in average temperature. To achieve this goal, the agreement promotes sustainability and directs investments towards environmentally positive impact solutions. In 2020, the European Union signed a similar agreement, known as the European Green Deal, which aims to promote a green transformation and limits the impact of human activities on the planet, through a series of coordinated actions at European level. In addition to these previous infinitives, the Global Reporting Initiative has defined a series of sustainable reporting standards suitable for any type of organization of different sectors.

Based on such initiatives and frameworks, more and more companies have developed integrated reports in which, in addition to financial information, environmental and social performance, and information are also disclosed (de Villiers et al., 2014).

Within the literature regarding non-financial reporting, there is a lack of studies concerning the public sector. While voluntary sustainability reporting in the private sector (Adams et al., 2014), mostly aligned with the Global Reporting Initiative (GRI) guidelines, is growing, there are little information about non-financial reporting practices in the public sector organisations.

Regarding the relationship between the results of non-financial reporting and stakeholders' engagement, there have been multiple studies on the private sector, while less on the public. Some studies, such as the one carried out by Kumar et al. (2021), have shown how cooperation between companies can be considered a strategy to improve performance in the economic, social, and environmental dimensions of sustainability. Other studies, such as the one conducted by Munjal and Sharma (2021), have focused on the sustainability report adopted by banks. Specifically, the research has revealed that stakeholder engagement in socially responsible practices strengthens the relationship between stakeholders and banks, which, in the end, improves financial performance.

3. Methodology

The paper conducts a Bibliometric analysis which consists of a hybrid methodology to facilitate a rigorous qualitative and quantitative examination of studied sources, combining bibliometric and code analysis (Cobo et al., 2011a; Massaro et al., 2016). This method based on the structured literature review (SRL) that requires sequenced and replicable processes of analysis (Tranfield et al., 2003). According to Secinaro et al. (2021) and Zupic and Čater (2015), the bibliometric approach includes a series of variables such as years, documents' information, sources, authors, keywords, citations, and countries. Hence, bibliometric analysis is a methodology that focus on measuring and analysing scientific production, dissemination, and the impact of research, with the main objective of providing an objective assessment of the quality and impact of research and its authors. It can include the quantification of publications, citations, collaboration among researchers and thematic trends in a specific field of study (Zupic and Čater, 2015). As stated in the contribution offered by Aria and Cuccurullom (2017) the tools and the process of scientific mapping have been described by the authors respectively in 2003 and 2011 (Börner et al., 2003; Cobo et al., 2011). The research was carried out through the SLR methodology, which allows for a systematic review of the literature on the subject through bibliometric analysis with qualitative and quantitative variables (Massaro et al., 2016). The analysis inspired by Zupic and Čater (2015) includes the following five phases:

The first step consists of defining the three research questions as previously mentioned in the introduction section (Massaro et al., 2016):

Then, it would be essential to write a research protocol containing the definition of information sources, the methods adopted, and the means and tools used to test and synthesise the studies (Petticrew and Roberts, 2006). Table 1 illustrates the review protocol and the different steps into which it is divided, including the main topic, the motivation of the choice, dataset creation, and the tools used to implement the analysis.

Table 1 Structured literature review protocol.

Research Protocol Elements	*Author's Consideration*
What is already known?	In the literature, there are several studies on how the private sector, by undertaking socially responsible practices and strengthening relationships with stakeholders, improves financial performance.
Motivation	To face stakeholder engagement and crisis era during which the innovation of the public sector involving all citizens and other stakeholders is fundamental.
Research topic	The study is based on the responsibility of the public sector to safeguard and increase public value, ensuring transparency and sustainable development that is socially inclusive and 'smart'. RQ1: What are the sources, keywords, social structures, and conceptual frameworks related to Smart Cities reporting practices? RQ2: What are the main theoretical and practical implications of the research? RQ3: What are the research limitations and possible future research directions of the topic analysed?
Journals' research	It was decided not to limit the research to individual scientific journals because of the still-young purpose. The analysis included articles and peer-reviewed journals. Additionally, the study selected "business, management and accounting" and "economics, econometrics and finance" sources from Scopus database to extract all relevant information sources.

Note: Authors' elaboration.

The third phase involves the description of the strategy and changes in the selection criteria. The bibliometric method used in this study can help readers to identify the main variables of the research flow (Secinaro et al., 2021). The study is based on the multidisciplinary database Scopus. This research database includes indexed and classified articles from the Institute for Scientific Information (Oakleaf, 2010). According to Okoli and Schabram (2010), to produce relevant sources, a fundamental step is conducting a keywords' analysis to investigate appropriate scientific information about Smart Cities reporting practices (Bamel et al., 2021; Bedford, 2015; Fteimi and Lehner, 2018). Hence, we used the following keywords research string within the Scopus database:

(TITLE-ABS-KEY ("public sector") OR TITLE-ABS-KEY ("public administration") OR TITLE-ABS-KEY ("public government") AND TITLE-ABS-KEY ("social") OR TITLE-ABS-KEY ("green") OR TITLE-ABS-KEY ("esg") OR TITLE-ABS-KEY ("citizen") OR TITLE-ABS-KEY ("sdg") OR TITLE-ABS-KEY ("public engagement") OR TITLE-ABS-KEY ("responsibility") OR TITLE-ABS-KEY ("public participation") AND TITLE-ABS-KEY ("smart city") OR TITLE-ABS-KEY ("gri") OR TITLE-ABS-KEY ("sustainability")) AND (LIMIT-TO (DOCTYPE , "ar") OR LIMIT-TO (DOCTYPE, "cp")) AND (LIMIT-TO (SUBJAREA, "BUSI") OR LIMIT-TO (SUBJAREA, "ECON")) AND (LIMIT-TO (LANGUAGE, "English")).

The analysis is based on an evolving field of research. Therefore, no time restrictions have been applied. In addition, the focus is based on database sources such as selected articles and conference papers in "Business, Management and Accounting" and "Economics, Econometrics and Finance", as the research questions are related to these two fields of study (Levy and Ellis, 2006). To increase the comprehensibility and quality of publications, only English-language articles (peer-reviewed) were considered (Massaro et al., 2016; Secinaro and Calandra, 2020). Applying the previous search criteria, 306 articles were selected. After defining the field of study, selecting the appropriate database, and identifying the search criteria, the fourth step related to the analysis tools that took over. Specifically, the analysis was performed using Bibliometrix, a statistical package available on R-Studio (Aria and Cuccurullo, 2017). The software retrieved relevant qualitative and quantitative bibliometric data such as authors, citations, country of production, or keywords. Finally, the fifth phase analyses the information and discusses the research results. Table 2 shows the method adopted during the research analysis.

Table 2 Features of the Bibliometric Study.

Research Steps	*Actions*
Step 1: Definition of the research field	Public sector with focus on socio-environmental commitment.
Step 2: Dataset selection	Scopus database.
Step 3: Selection criteria modifications	Keywords: Public Sector, Public Administration, Responsibility, Sustainability, Green, Social, Citizen, ESG, SDGs, Smart City, Public Engagement, Public Participation. No time restriction. Only articles and conference papers.
Step 4: Analysis tools	Bibliometrix, R package.
Step 5: Information and data analysis	Results analysis and discussion.

Note: Authors' elaboration.

4. Results

Firstly, the study starts by describing the characteristics of the dataset and the related implications of the 306 results. The analysis of these results is focused on understanding the evolution of the literature over the years, highlighting data trends and describing current and previous research to suggest future research lines. Therefore, in the next paragraphs, research information is presented, which includes relevant sources, keyword analysis, relevant data on authors, and social, thematic, and intellectual structure.

Table 3 Main information.

Data Source	*Description*	*Value*
Timespan	Publication years	1992–2002
Sources (journals, books, etc.)	Distribution frequency	197
Documents (papers)	Total number of documents	306
Annual growth rate (%)	–	12
Document Average Age	Average number of citations for each document	5,4
Average citations per doc	Average number of annual citations in each document	2,06
References	Total number of references	23,027
Document contents		
Keywords Plus (ID)	Total number of words appearing frequently in the article, titles and references	968
Author's Keywords (DE)	Total number of keywords cited by the author	1,210
Authors		
Authors	Total number of authors	817
Authors of single-authored docs	Number of single-author document	64
Authors collaboration		
Single-authored docs	–	94
Co-Authors per doc	–	3,27
International co-authorships %	–	35,6
Document type		
Article	–	306
Review	–	30

Note: Authors' elaboration.

5. Main Information

The collected articles were published over a period of time from 1992–2022. Only from the biennium of 2014/2015 did the number of publications begin to increase, even becoming more substantial (Figure 1). One reason for this increase may be that since that biennium, there has been a consistent interest in sustainability from various international and scientific organizations in all its socio-environmental forms. Confirming this, in 2015, all member states of the United Nations adopted the 2030 Agenda for Sustainable Development, based on the 17 Sustainable Development Goals (SDGs). Also, in the same year, 183 members of the United Nations Convention on Climate Change (UNFCCC) signed on to the Paris Agreement, an international treaty that incentivises climate resilience by limiting global temperature and cohesively directing investments to the path of environmental improvement. These events help explain the increase in the number of sustainability publications. It can also be noted that there has been a decline in publications in recent years, and this may be because the whole world has shifted its attention to the COVID-19 pandemic that occurred in 2020, setting aside the focus on research issues for the period thereafter.

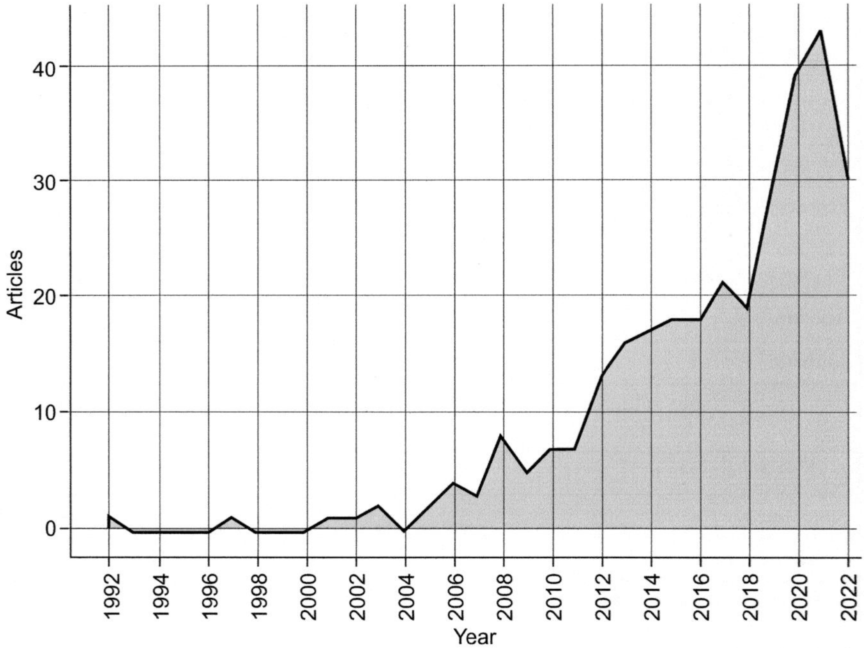

Fig. 1 Published articles distribution.
Source: Biblioshiny.

As shown in Figure 2, citations have not followed the same trend as publications. In particular, the line graph shows an irregular trend. Moreover, these results suggest poor productivity in recent years.

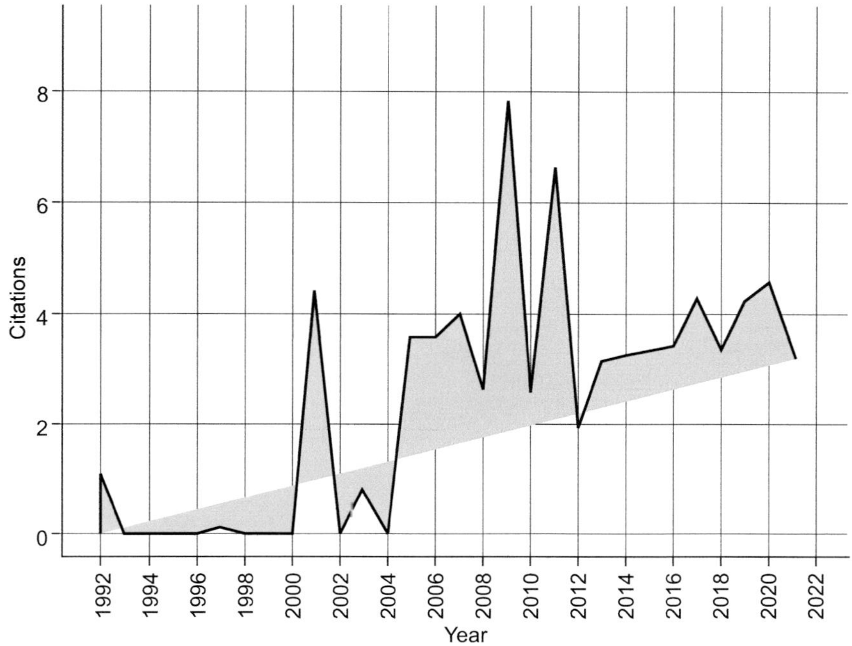

Fig. 2 Citations' distribution.
Source: Biblioshiny.

6. Relevant Sources

The top 10 sources correspond to about 20% of scientific production. Table 4 shows *Journal of Cleaner Production* in the first place, not only by several articles but also by several citations, amounting to 23 and 321, respectively. In second place is the *Public Money and Management*, with six published articles, followed by the *Journal of Sustainable Tourism* and the *Sustainability Accounting, Management and Policy Journal*, both with five published articles. As for citations, in second place, with only four publications, is the *Journal of Public Budgeting, Accounting and Financial Management*, which has 246 citations, followed by the *Journal of Sustainable Tourism*, with 126 citations. As for the *Journal of Cleaner Production*, (Figueira et al., 2018; Giacomini et al., 2021; Walker and Preuss, 2008) the publications analysed adopting sustainability policies and practices in public sector organisations. The *Public Money and Management* publications (Montecalvo et al., 2018) referred to the link between sustainability, citizen involvement in the social innovation process, and value creation from the public sector perspective.

7. Thematic Map

In this paragraph, information related to the authors' keywords regarding the relationship between public administrations and socio-environmental principles is

Table 4 Most relevant sources.

Sources	*Articles*	*Citations*
Journal of Cleaner Production	23	321
Public Money and Management	6	75
Journal of Sustainable Tourism	5	126
Sustainability Accounting, Management and Policy Journal	5	50
Journal of Public Budgeting, Accounting and Financial Management	4	246
Cities	4	45
Meditari Accountancy Research	4	37
Technological Forecasting and Social Change	4	22
Business Strategy and the Environment	3	73
American Review of Public Administration	3	34

Note: Authors' elaboration.

presented. Figure 3 represents the Word Cloud of keywords used by the authors. This figure includes keywords such as "public sector", 'sustainability', "smart city", 'decision-making', and "governance approach". Generally, the term "Public Sector" refers to a set of institutions that manage the interests of multiple subjects requiring various public activities and services in different ways. The concept of 'sustainability' began to appear in the academic lexicon around the mid-1980s (Secretary-General and Development, 1987), and since then, it has undergone substantial evolution. Following the birth of the sustainability term, the concept of corporate social responsibility became part of the academic lexicon, and more and more organisations started to adopt specific policies to maximise a company's profits through actions that simultaneously respect environmental and social needs. Subsequently, the term 'sustainability', which originated with a definition focused strictly on ecological themes and aspects, has expanded to also encompass aspects concerning the social (S) and governance (G) dimensions, in addition to the environmental (E). The first two keywords are closely connected as the public sector plays a crucial role in promoting and implementing sustainability. Especially amid strong environmental and social changes, ESG factors increasingly impact public finance and decision-making processes (Zioło et al., 2022). Specifically, the public sector can act through sustainable policies and programs, funding social projects, and raising awareness on environmental issues. Public organisations increasingly consider environmental and social aspects in their implementation strategies (Gelderman et al., 2017). Furthermore, the public sector can lead in adopting sustainable practices, inspiring other sectors to follow suit. Sustainability, in turn, contributes to the well-being of society and the achievement of sustainable development goals, improving the quality of life for all citizens. In this regard, the need to pursue sustainable development has become a central theme; this is also due to citizens' increased sensitivity to improving environmental and social conditions (Castellano et al., 2022). The connection between the first two keywords is the subject of this study. Specifically, the research concerns the decision-making process of public administrations regarding social and environmental innovation projects directed at interest groups regarding the inclusion and participation of local citizens. Indeed, as stated by Khozhylo et al. (2022) referring to the decision-making process, public management should be considered an intersectoral complex formed through the involvement of local entities, business partners, and citizens to create, maintain, and develop a favourable and sustainable environment. Public sector innovation is fundamental for the development of "Smart Cities", another keyword (Ang-Tan and Ang, 2022). A Smart City, or intelligent city, is a model of organisation that has developed thanks to the progressive digital and ecological transition. In recent global urban issues such as climate change and energy problems, the Smart City has been proposed as one of the solutions in urban planning (Choi and Siqin, 2022). Indeed, following the demographic development increasingly concentrated towards cities, the necessity of implementing urban efficiency and smart cities is even more of a prerogative. According to the European Union, "Smart Cities" includes six main dimensions. The first dimension identified for research purposes concerns the "smart environment". This last term means sustainable development, reduced ecological footprint, and energy efficiency. Governments worldwide are increasingly embarking on the path of "green growth" (Capasso et al., 2019). With the growing interest in sustainability, public organisations are increasingly adding strategic objectives related to environmental and social aspects (Gelderman et al., 2017). Another relevant dimension to the subject of study is "smart governance", which means that human capital, sustainability, and community must be considered. The relevance of collaborative governance models in smart cities is fundamental to promoting citizen involvement in the city's public affairs (Bolívar, 2018). Only collaboration between the public sector, civil society organisations, and the community will lead to successful inclusive innovation (Farneti and Siboni, 2011). Another significant dimension is "Smart People", which refers to the involvement and participation of citizens in the public decision-making process through a bottom-up approach and participatory policy. Civic participation, in addition to being a fundamental goal within the context of the Smart City, is the main way to transform the government, making it more open and closer to the needs of the citizens (Bolívar, 2018). In recent years, numerous documents containing non-financial information have arisen with the main objective of satisfying the interest and involvement of stakeholders (Montesinos and Brusca, 2019). To manage the environment in a balanced, sustainable, and responsible manner, more systematic reporting at the aggregate level by governments is essential (Johnson et al., 2002). With particular regard for the public sector, a useful tool to increase the level of information and transparency of citizens towards public administrations can be represented by the *Popular Financial Report* (Montesinos and Brusca, 2019). It is a document used to make the budget of public entities understandable to citizens, thus increasing transparency and reliability. Budget transparency could be positively associated with the financial sustainability practices of governments and with the objectives of increasing the trust and participation of citizens (Cuadrado-Ballesteros and Bisogno, 2022). In this regard, transparency is associated with the financial sustainability practices of governments as it allows citizens, investors, and stakeholders to understand how public finances and resources are managed. Financial transparency also increases trust in public institutions, attracts investments, and contributes to more efficient management of public finances.

Fig. 3 WordCloud.
Source: Biblioshiny.

The TreeMap below (Figure 4) highlights the combination of possible keywords within the research. It can be noted that the most important terms in the Figure are "Sustainable Development", "Public Sector", and "Public Administration", followed by "Decision Making" and "Smart City". The order and arrangement of the words are not random; in fact, it can be seen that the term with the highest number of results is "sustainable development", amounting to 14%, and "public sector" and "public administration", totalling 16%. Part of the research is indeed represented by the concept of sustainable development, which, as previously mentioned, encompasses concepts that go beyond purely environmental themes and include social and governance aspects. Following a growing interest in sustainability issues, public organisations are increasingly considering information related to environmental and social aspects among their strategic objectives (Gelderman et al., 2017). Furthermore, to design effective policies, it is necessary to generate governance types geared towards greater efficiency and greater social cohesion (Vara-Sánchez et al., 2021). Therefore, there is a strong connection between sustainable development practices and the actions

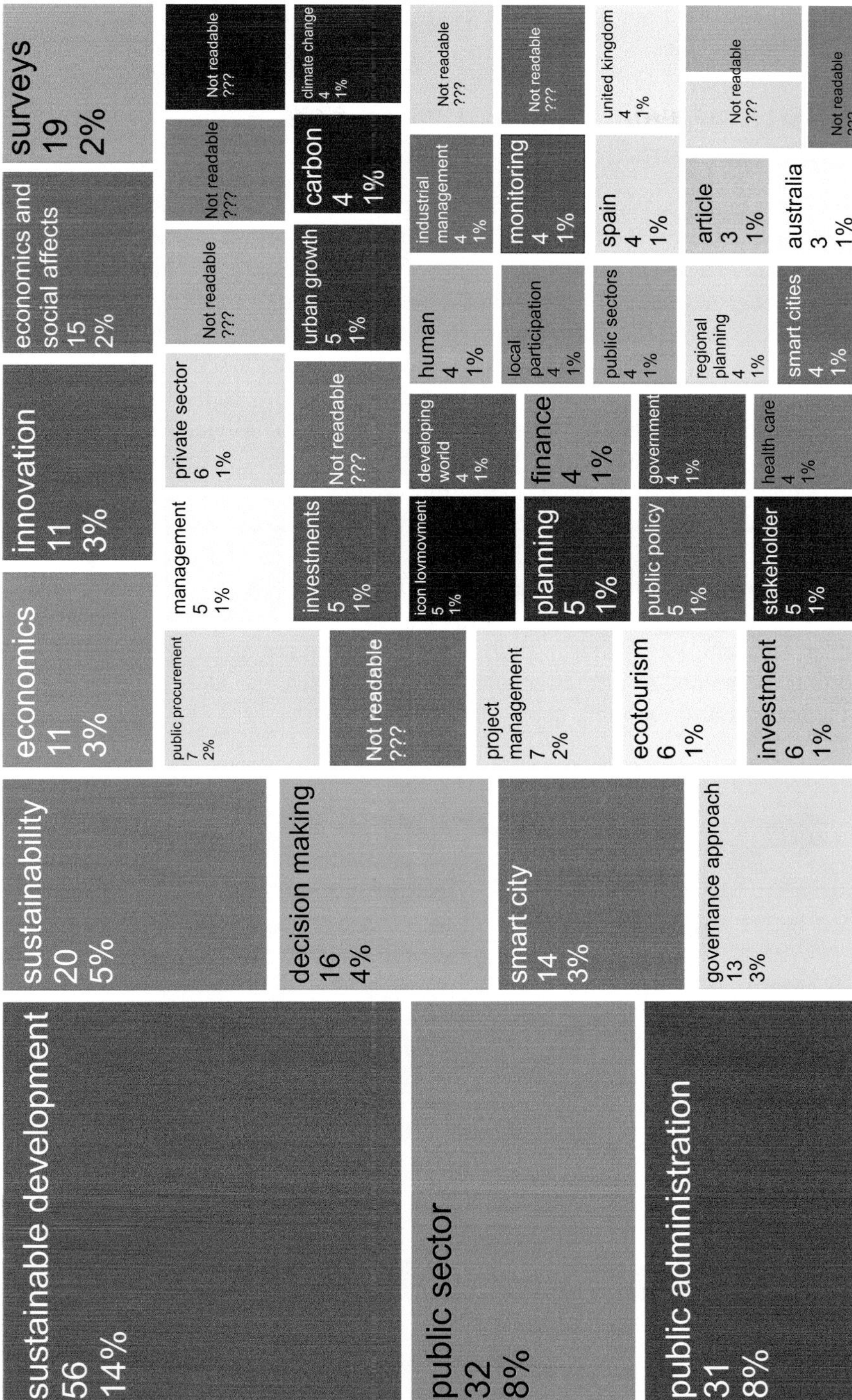

Fig. 4 TreeMap.
Source: Biblioshiny.

of the public sector, both from the perspective of social responsibility, inclusion, and investment choices.

8. Social Structure (Countries, Institutions, and Relevant Authors)

As shown in Figure 5, various countries have published articles about this specific research subject. However, most of the published articles are concentrated in a few countries. There has been a significant presence of European countries. This may be due to a developed financial/banking market. Moreover, as expected, there has also been a broad presence of English-speaking countries (such as the United Kingdom, Australia, and the USA). Regarding Europe, the first country with the highest number of works is Italy, with 106 articles. It is followed by Spain, with 65 publications, and Portugal, in 10th place, with 23 published articles. As mentioned earlier, there is a strong presence of English-speaking countries; the United States is in second place, with 86 publications, followed by the UK, on par with Spain, with 65 articles, and Australia, in 5th place, with 58 published works. The major presence of these countries is because of the use of documents containing non-financial information, with particular regard for the POP Budget, which represents the subject of study and originated in the United States, generally referred to as Popular Financial Reporting. Indeed, in the United States and Canada, popular reporting understood as information aimed at the citizen, has already taken root. As the authors also reported, about 75% of local entities in the country prepare

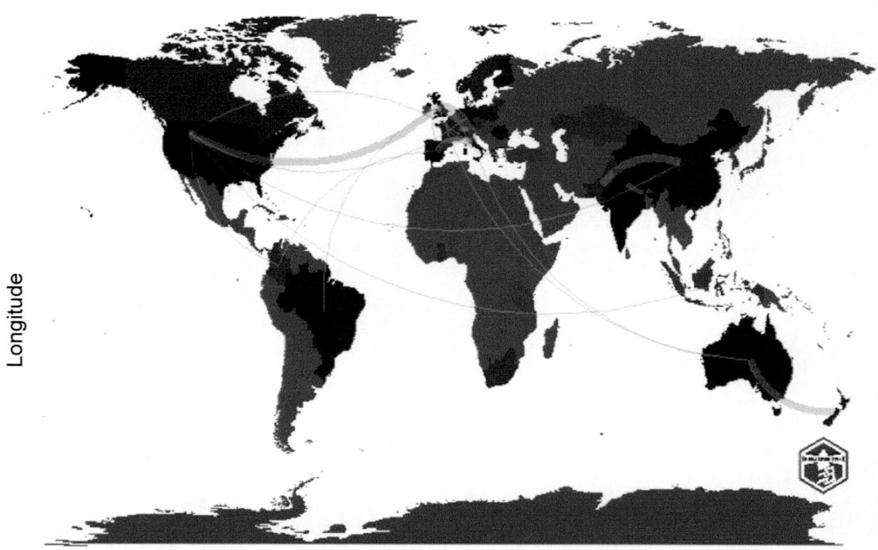

Fig. 5 WorldMap.
Source: Biblioshiny.

them periodically (Montesinos and Brusca, 2019). Regarding the countries with the most cited articles, the results are not much different from those found previously. In particular, the UK is in the first place with 1,775 citations, followed by Italy with 468, Austria with 323, and Spain with 259 citations. The results are also reflected in the affiliations: the most relevant have been Rensselaer Polytechnic Institute, with 18 articles published; not reported, with 12 publications; and Cardiff University and the University of Parma, with 8 articles each. Finally, considering the output of the affiliations over time, it turns out that the publication of the authors' scientific production has been active only since 2003, so a particular author who has contributed to the development of the subject throughout the entire period under analysis cannot be identified.

9. Conceptual Structure

The conceptual structure identifies and visualises the topics and thematic areas of the field of research (Cobo et al., 2011). Therefore, this section includes an analysis of thematically related trending topics in this field and the dendrogram of topics (Secinaro et al., 2021). Figure 6 defines trending topics based on Keywords Plus, which are words or phrases frequently appearing in article references' titles but not in the article title itself. Keywords Plus is developed with an algorithm that enhances the search power of cited references by looking across all disciplines for articles with commonly cited references (Secinaro et al., 2021). Keywords Plus can also be presented for articles that do not have author keywords or may include important terms not listed among the author keywords (Secinaro et al., 2021). Figure 6 highlights that, during the analysed period, there has been a change in the trend of topics. The year 2014 saw the greatest change, particularly with increased studies on "innovation", "sustainability", and "economic and social effects". As mentioned earlier, this change may be because in the following year, the United Nations member states introduced the 2030 Agenda for Sustainable Development, and the Paris Agreement was signed, in addition to the fact that from that year, states and the scientific world have focused more on the study of sustainability. Another significant change occurred in 2016 when research focused on the "governance approach", "public administrations", and "environmental protection". Finally, in the period from 2018–2022, in addition to the intensification of studies on already known concepts, such as "sustainability" and "public sector decision-making", there is the presence of terms, such as "Smart City" and "sustainable development goals", demonstrating that in recent years, there has been a more rigorous concentration about the study in question. Although there is no single moment to determine when a city becomes smart, many cities have gradually implemented "smart" solutions during the considered period. Moreover, the number of smart cities has likely increased significantly in this period due to the growing focus on sustainable and technological solutions to address the challenges of modern cities.

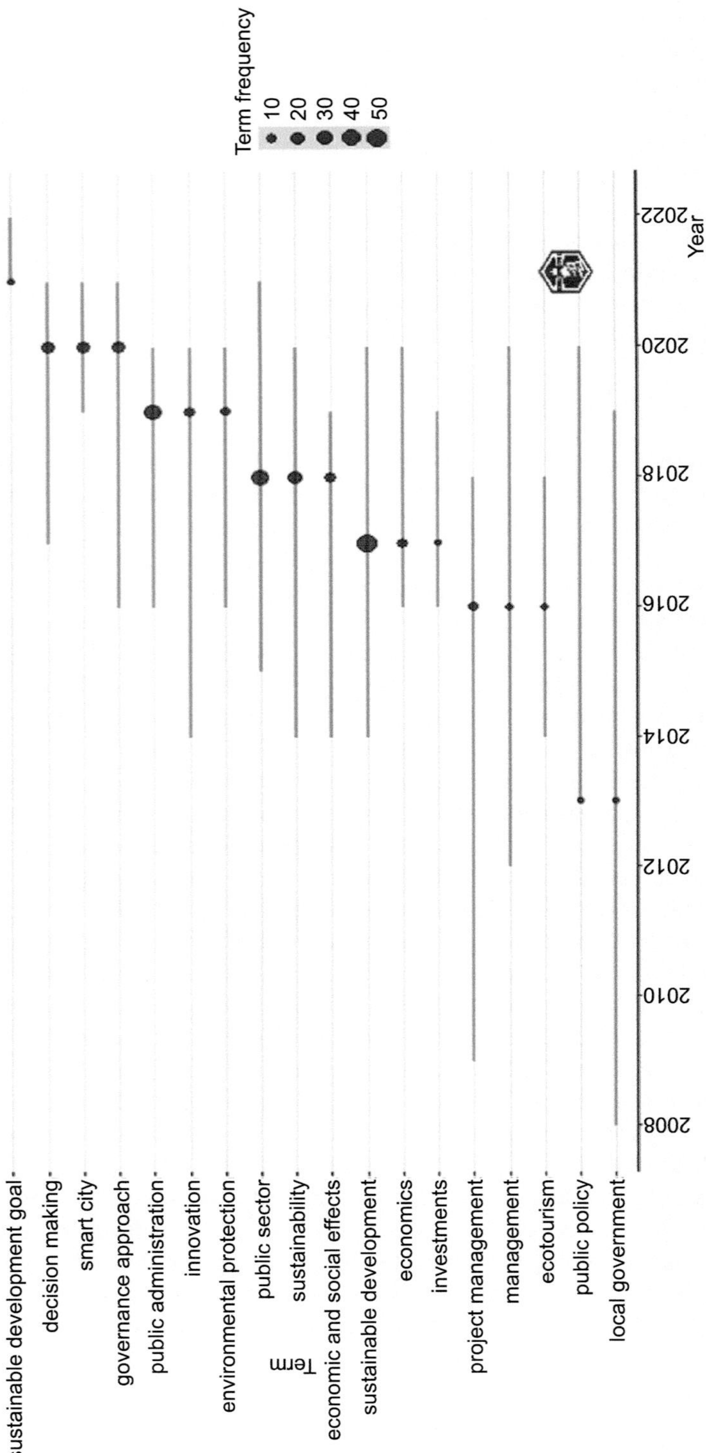

Fig. 6 Trend topic.
Source: Biblioshiny.

Table 5 highlights the total number of Keywords Plus in the top 10 positions.

Table 5 Top 10 Keywords Plus.

Keywords	*No. of Articles*
Sustainable Development	56
Public Sector	32
Public Administration	31
Sustainability	20
Decision-making	16
Smart City	14
Governance Approach	13
Economy	11
Innovation	11
Economic and Social Effects	10

Note: Authors' elaboration.

Figure 7 illustrates a thematic map that shows each keyword in the two-dimensional space explained by the axes (Noyons, 2001). The scale of the points is proportional to the cumulative input of each keyword, and by applying a clustering algorithm to a network of keywords, it is possible to highlight the various themes of a given domain (Secinaro et al., 2021). Each cluster can be represented on a special chart, known as a strategic or thematic map, where centrality can be read as the importance of the theme in the field considered, while density can be read as a measure of the theme's growth (Secinaro et al., 2021). Observing Figure 7, the niche themes, therefore intended for a specific field, concern "project management" and "industrial management". As basic and necessary themes, terms such as "sustainability" and "governance approach" have been detected. Furthermore, the driving themes in Figure 7 are "sustainable development", "public administration", "innovation", "local participation", and "smart city". Finally, the declining themes concern "health policy" and "knowledge".

Figure 8 shows a dendrogram to observe the hierarchical relationship between keywords (Forina et al., 2002). The diagram shows four relative analysis lines: red, blue, green, and purple. This figure visualises the similarity in the "clustering" process in which each subdivision represents relevant themes and keywords (Secinaro et al., 2021).

Regarding the blue cluster, terms such as 'sustainability", "governance approach", and "sustainable development goals" have been highlighted. To design effective policies, it is necessary to generate types of governance geared towards greater efficiency and social cohesion (Vara-Sánchez et al., 2021). Indeed, with the growing interest in sustainability, among the strategies and objectives of public organisations, more and more information related to environmental and social aspects is being considered, particularly regarding the concept of corporate

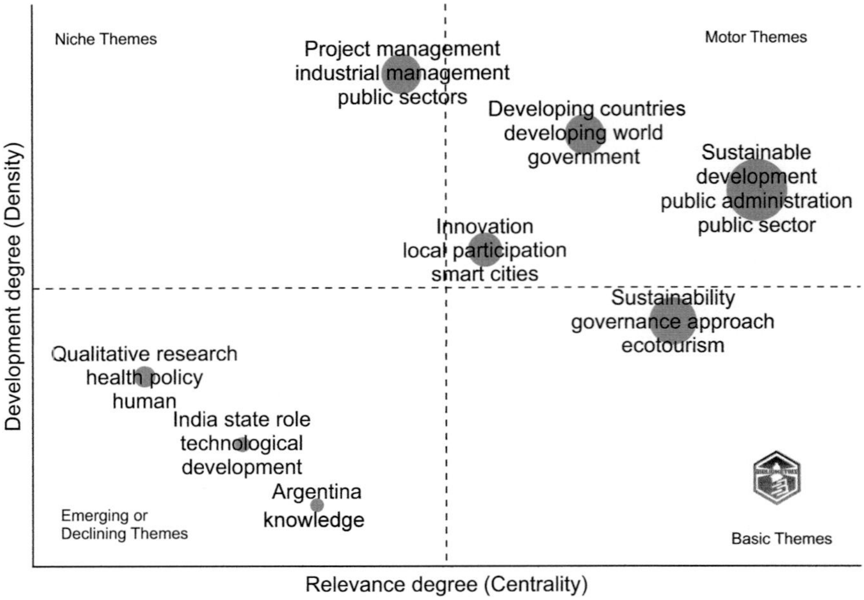

Fig. 7 Thematic map.
Source: Biblioshiny.

social responsibility (CSR) (Gelderman et al., 2017). Regarding the SDGs, some authors have stated that the 17 United Nations sustainable development goals have created a framework for environmental and social impacts that institutions and societies use to guide resource allocation (Consolandi et al., 2020). From this grouping of words, it emerges that public administration is a tool aimed at improving the well-being of people and the environment through entrepreneurial actions and projects. Sustainability, SDGs, and the governance of public entities are linked and are interdependent elements. Good governance of public entities is essential to achieve the SDGs and to ensure sustainable development at a global level. The second cluster under evaluation is that of the green strand. The latter, coming very close conceptually to the former, contains words such as "public policy" and "environmental protection". Green growth is seen as an effective way to achieve environmental sustainability, and institutions actions, such as, for example, incentives for education and research and development, contribute significantly to enhancing green economic growth (Huang et al., 2018). The third cluster, which contains the highest number of words, is represented by purple. Terms such as "urban growth", "smart city", "climate change", and "public sector" can be noted. As reported Alim and Polak (2016), the public sector is responsible for safeguarding the public interest and ensuring transparency, social inclusion, and "smart" sustainable development. Furthermore, insights into "smart cities" and the literature on public innovation have also been used to identify governments' internal and external collaboration capabilities that should lead to innovation in the

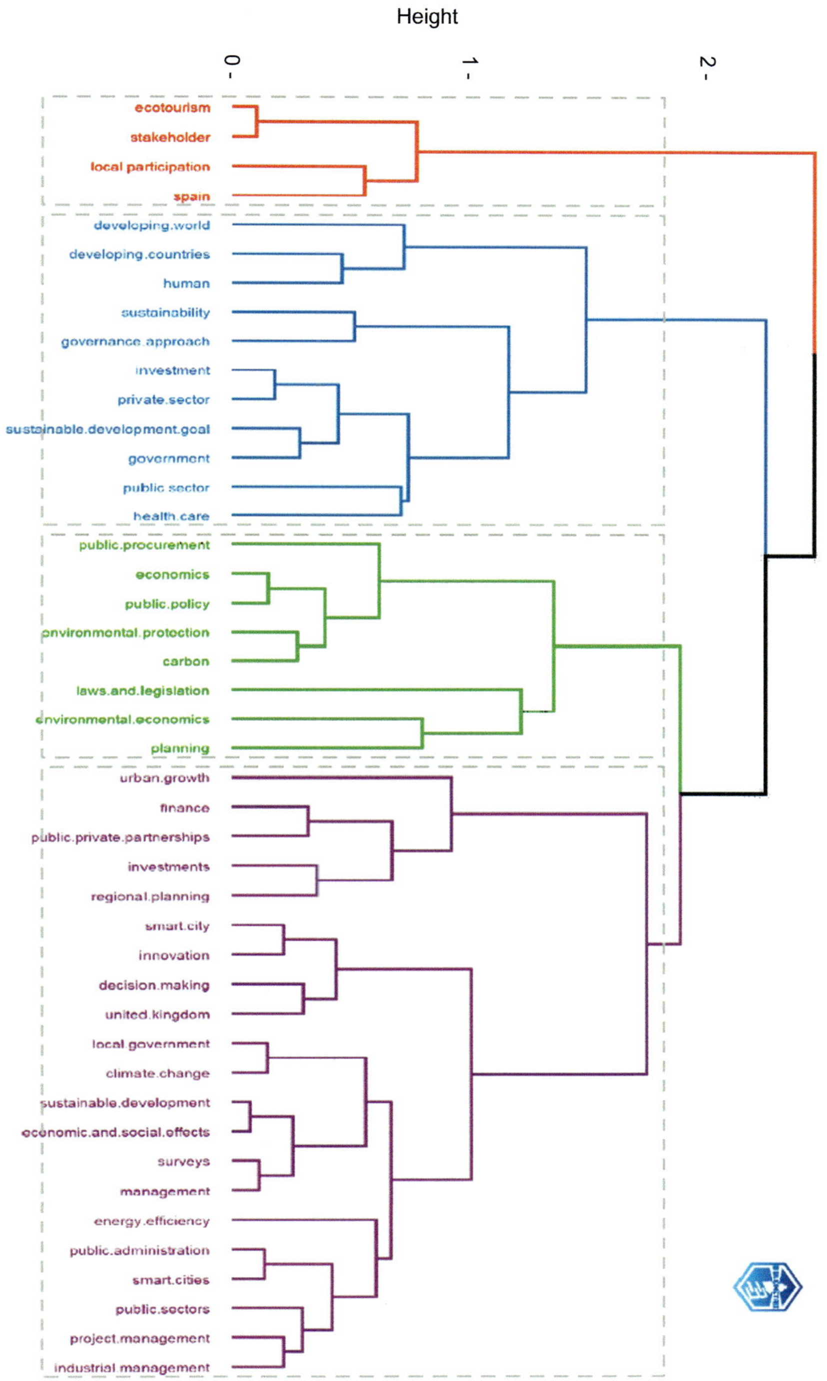

Fig. 8 Topic Dendrogram.
Source: Biblioshiny.

public sector (Barrutia et al., 2022). Finally, we analyse the last cluster represented by the colour red. The terms 'stakeholders' and "local participation" denote the importance of citizen involvement in innovation and socio-environmental development under public administrations. Civic participation is the main way to transform the government, making it more open and closer to the needs of the citizens, as well as being a principal objective within the context of Smart Cities (Bolívar, 2018). In this regard, as reported Montesinos and Brusca (2019), several non-financial documents are available to public administrations to increase transparency and engagement towards citizens. As previously mentioned, one of these documents is the PFR.

Figure 9 aims to outline the results of the analysis of citations and co-citations (Secinaro et al., 2021). The author's co-citation analysis offers a range of perspectives to consider the effect of these scholars on the academic framework of the public sector and sustainability. The size of each cluster indicates the degree of co-citation of the authors, and the width of the lines shows the strength of

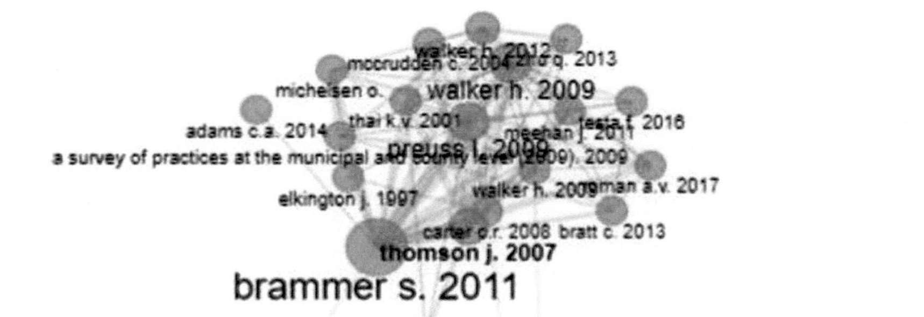

Fig. 9 Citation and co-citations analysis.
Source: Biblioshiny.

the co-citation relationships, while the proximity between authors is seen from the connections, not the spatial proximity between them (Secinaro et al., 2021). Finally, the edges of the network can have different meanings based on the type of citation (co-citation or direct citation) (Secinaro et al., 2021). From the following diagram, it is possible to classify four clusters of different colours: green, red, blue, and purple. The purple subgroup, consisting of only four co-citations, has been defined as an outlier. Above the centroid, the first cluster, marked in green, can be described as focusing on sustainability in the public sector. The red cluster consists of research on sustainability reports. Finally, the theme in blue focuses on the strategic decision-making process of public administrations.

10. Discussion

This bibliometric analysis focused on researching the decision-making process of public administrations toward socio-environmental principles, focusing on public value creation and stakeholder engagement. The study investigated qualitative and quantitative variables such as authors, citations, journals, and topics. Among them, some important observations emerged. For example, Farneti F. and Walker H. are the authors with the largest publications. According to Walker and Preuss (2008), the need to reduce the social and environmental footprint to stimulate sustainability leads to the development of new policies at government and organisational levels (Wontner et al., 2020). Furthermore, public organisations increasingly include information related to environmental and social aspects among their strategic objectives (Gelderman et al., 2017). Focusing on non-financial information, some of Farneti's contributions have concentrated on the applicability of the Global Reporting Initiative guidelines and sustainability reporting to public sector organisations (Farneti et al., 2019; Farneti and Siboni, 2011). In the document *Guidelines for Sustainability Reporting* there is specific information regarding stakeholder engagement. Their involvement concerns the preparation of the sustainability report and a guide based on stakeholders' expectations in defining sustainability strategies that adapt to their development models. Therefore, a certain trend line can be noted among many authors considered in the research regarding the importance of the link between the public sector and its actions, sustainability, understood in its broadest sense, and stakeholder involvement in citizen engagement. Citizen-oriented management, i.e., internal management focused on innovation combined with strong collaboration with citizens, seems to have a positive influence on public value and, therefore, on efficiency, effectiveness, and the ability to address the social challenges of the sector (Barrutia et al., 2022). Expanding the topic, the public sector should formulate specific recommendations to create business sectors that are socially inclusive and sustainable and that incorporate "smart technology" (Crick, 2022). In this regard, for the development of smart cities, public sector innovation represents a fundamental step (Ang-Tan and Ang, 2022). Additionally, Bolívar (2018) addressed the theme of smart cities, stating that collaborative governance models in smart cities aim to improve citizen participation in the public arena. Therefore, the public innovation process implies a series of factors beyond creating a green-based economy. Together with the cooperation of citizens and the community, the government might succeed in addressing all sustainable development issues for society and the environment. In practice, to manage the environment in a balanced, sustainable, and accountable way, more systematic reporting at the aggregate level by governments is essential (Johnson et al., 2002). Several non-financial documents are available to governments in the literature to increase transparency and engagement with citizens (Montesinos and Brusca, 2019). One such document, which is part of the object of study, is the PFR. This document's function is to make readable a complex accounting tool that represents the performance of local public bodies integrated with the accounting data of companies and organisations delegated in the provision of services and benefits.

Continuing the discussion of the research, the work shows that the journals with the highest number of articles on these topics have been the *Journal of Cleaner Production*, with 23 publications and *Public Money and Management*, with six publications. Regarding the *Journal of Cleaner Production*, publications analysed the adoption of sustainability policies and practices in public sector organisations (Figueira et al., 2018; Giacomini et al., 2021; Walker and Preuss, 2008). Meanwhile, publications from *Public Money and Management* have referred to the link between sustainability, citizen involvement in the process of social innovation, and value creation from the perspective of the public sector (Montecalvo et al., 2018). Following the published articles, keyword analysis also revealed some points of similar interest. Terms such as "public sector", 'sustainability', "smart city", "decision making", and "governance approach" emerge. In this case, a common thread can be noted regarding the effect of the decision-making process of public administrations on innovation projects for the environment and interest groups, particularly regarding local citizens.

Conclusion

In conclusion, this chapter investigates the connection between public administrations, socio-environmental parameters, and citizen involvement. A bibliometric analysis was conducted using the R Bibliometrix package to answer these research questions. Building on the literature of the investigated topics, this work provides a qualitative and quantitative analysis of bibliometric variables. The study demonstrates that the intervention of the public sector in favour of sustainable development promotes societal innovation, increases public value, and creates a more inclusive and participatory community, leading to smart city contexts. The research is based on extensive scientific literature showing how the public sector's innovation process and sustainable development are conceptually related. In recent years, the term "sustainable development" has expanded beyond environmental protection and ecological footprint reduction to include social issues and good governance. The analysis also offers practical suggestions, particularly regarding reporting. Financial transparency can positively impact the financial sustainability of governments, in addition to increasing citizen trust and participation. One of the reporting tools used by public administrations is the PFR. It seeks to address recent challenges that are emerging with increasing relevance by responding to the need to foster transparency and incisiveness in the operation of the public sphere, trying not to eliminate or surpass differences among various groups (Biancone et al., 2019).

The paper, like any other research, has some limitations due to the novelty of the topic, which is becoming extremely relevant in recent years. The first limitation can be found in the use of only the Scopus database to perform a bibliometric analysis of the literature. Scopus database provides a robust platform for academic insights, but it may inadvertently omit significant literature and overlooked some publications and other documents relevant to the research not indexed in its system which, on the other hand, could be covered by other databases. Secondly the research considers only peer-reviewed articles and reviews written in English in business, management, and accounting fields and on those of economics, econometrics, and finance which may not fully encompass the interdisciplinary nature of smart cities. In addition, the study has focused the attention on a limited number of articles in the sample to ensure rigorous methodology, which may exclude important studies on the topic under analysis.

Simultaneously, scholars may consider this study a valuable resource for understanding the state of the art of this specific field of analysis and identify potential future research lines such as, for example, the examination of the use of technology in the public sector sphere. An interesting insight could be the more detailed analysis of the role of Internet of Things (IoT) technology in managing urban infrastructures, exploring how IoT can be effectively integrated within urban planning to provide more responsive and sustainable city environments. Furthermore, an exploration of digital transformation policies within the public sector could shed light on the process of policy formulation, adoption, and its eventual outcomes on technological uptake. Another relevant aspect that could be investigated in future research is that of the dynamics of public engagement through digital platforms. How these platforms are shaping public administration activities, and the way social media can be used to capture public opinion. Additionally, comparative studies could provide a broader perspective, analysing how diverse cultural and policy environments influence the adoption and effectiveness of technology in the public arena. There is also an emerging need to scrutinize the long-term sustainability of technological innovations in the smart city context, looking beyond the initial implementation to the enduring socio-economic impacts these technologies have on various population segments.

References

Alim, S. and Polak, J. (2016). Public–private partnerships for future urban infrastructure. *In*: *Proceedings of the Institution of Civil Engineers - Management, Procurement and Law*, *169*(4), 150–158. https://doi.org/10.1680/jmapl.16.00001.

Ang-Tan, R. and Ang, S. (2022). Understanding the smart city race between Hong Kong and Singapore. *Public Money and Management*, *42*(4), 231–40. https://doi.org/10.1080/09540962.2021.1903752.

Aria, M. and Cuccurullo, C. (2017). Bibliometrix: An R-tool for comprehensive science mapping analysis. *Journal of Informetrics*, *11*(4), Articolo 4. https://doi.org/10.1016/j.joi.2017.08.007.

Artamonov, G.E., Erofeeva, V.V., Gutnikov, V.A. and Yablochnikov, S.L. (2020). Modeling the Environmental Situation in a Smart Citiy. *In*: *2020 International Conference on Engineering Management of Communication and Technology (EMCTECH)*, 1–4. https://doi.org/10.1109/EMCTECH49634.2020.9261566.

Bamel, N., Pereira, V., Bamel, U. and Cappiello, G. (2021). Knowledge management within a strategic alliances context: Past, present, and future. *Journal of Knowledge Management*, *25*(7), Articolo 7. https://doi.org/10.1108/JKM-06-2020-0443.

Barrutia, J.M., Echebarria, C., Aguado-Moralejo, I., Apaolaza-Ibáñez, V. and Hartmann, P. (2022). Leading smart city projects: Government dynamic capabilities and public value creation. *Technological Forecasting and Social Change*, *179*, 121679. https://doi.org/10.1016/j.techfore.2022.121679.

Bedford, D.A.D. (2015). Enhancing Access to Knowledge Management Literature—A Proposal for Domain-Based Classification Scheme and Thesaurus. *Journal of Information and Knowledge Management*, *14*(01), Articolo 01. https://doi.org/10.1142/S0219649215500069.

Biancone, P.P., Secinaro, S. and Brescia, V. (2016). The Popular Financial Reporting: Focus on Stakeholders: The first European Experience. *International Journal of Business and Management*, *11*(11), Articolo 11. https://doi.org/10.5539/ijbm.v11n11p115.

Biancone, P., Secinaro, S., Brescia, V. and Iannaci, D. (2019). The Popular Financial Reporting between Theory and Evidence. *International Business Research*, *12*(7), Articolo 7. https://doi.org/10.5539/ibr.v12n7p45.

Bolívar, M.P.R. (2018). Creative citizenship: The new wave for collaborative environments in smart cities. *Academia Revista Latinoamericana de Administración*, *31*(1), 277–302. https://doi.org/10.1108/ARLA-04-2017-0133.

Börner, K., Chen, C. and Boyack, K.W. (2003). Visualizing knowledge domains. *Annual Review of Information Science and Technology*, *37*(1), 179–255. https://doi.org/10.1002/aris.1440370106.

Brescia, V. (2019). *The Popular Financial Reporting: New Accounting Tool for Italian Municipalities*. 153. FrancoAngeli, Italy.

Capasso, M., Hansen, T., Heiberg, J., Klitkou, A. and Steen, M. (2019). Green growth: A synthesis of scientific findings. *Technological Forecasting and Social Change*, *146*, 390–402. https://doi.org/10.1016/j.techfore.2019.06.013.

Castellano, R., Punzo, G., Scandurra, G. and Thomas, A. (2022). Exploring antecedents of innovations for small- and medium-sized enterprises' environmental sustainability: An interpretative framework. *Business Strategy and the Environment*, *31*(4), 1730–48. https://doi.org/10.1002/bse.2980.

Choi, T.-M. and Siqin, T. (2022). Blockchain in logistics and production from Blockchain 1.0 to Blockchain 5.0: An intra-inter-organizational framework. *Transportation Research Part E: Logistics and Transportation Review*, *160*, 102653. https://doi.org/10.1016/j.tre.2022.102653.

Cobo, M.j., López-Herrera, A.g., Herrera-Viedma, E. and Herrera, F. (2011). Science mapping software tools: Review, analysis, and cooperative study among tools. *Journal of the American Society for Information Science and Technology*, *62*(7), Articolo 7. https://doi.org/10.1002/asi.21525.

Consolandi, C., Phadke, H., Hawley, J. and Eccles, R.G. (2020). Material ESG Outcomes and SDG Externalities: Evaluating the Health Care Sector's Contribution to the SDGs. *Organization and Environment*, *33*(4), 511–33. https://doi.org/10.1177/1086026619899795.

Crick, A.P. (2022). What innovations would enable the tourism and hospitality industry in the Caribbean to re-build? *Worldwide Hospitality and Tourism Themes*, *14*(6), 534–40. https://doi.org/10.1108/WHATT-05-2022-0060.

Cuadrado-Ballesteros, B. and Bisogno, M. (2022). Budget transparency and financial sustainability. *Journal of Public Budgeting, Accounting and Financial Management*, *34*(6), 210–34. https://doi.org/10.1108/JPBAFM-02-2022-0025.

Dal Mas, F., Massaro, M., Lombardi, R. and Garlatti, A. (2019). From output to outcome measures in the public sector: A structured literature review. *International Journal of Organizational Analysis*, *27*(5), 1631–56. https://doi.org/10.1108/IJOA-09-2018-1523.

Farneti, F., Guthrie, J. and Canetto, M. (2019). Social reports of an Italian provincial government: A longitudinal analysis. *Meditari Accountancy Research*, *27*(4), 580–612. https://doi.org/10.1108/MEDAR-11-2018-0397.

Farneti, F. and Siboni, B. (2011). An analysis of the Italian governmental guidelines and of the local governments' practices for social reports. *Sustainability Accounting, Management, and Policy Journal*, *2*(1), 101–125. https://doi.org/10.1108/20408021111162146.

Figueira, I., Domingues, A.R., Caeiro, S., Painho, M., Antunes, P., Santos, R., Videira, N., Walker, R.M., Huisingh, D. and Ramos, T.B. (2018). Sustainability policies and practices in public sector organisations: The case of the Portuguese Central Public Administration. *Journal of Cleaner Production*, *202*, 616–30. https://doi.org/10.1016/j.jclepro.2018.07.244.

Forina, M., Armanino, C. and Raggio, V. (2002). Clustering with dendrograms on interpretation variables. *Analytica Chimica Acta*, *454*(1), 13–19. https://doi.org/10.1016/S0003-2670(01)01517-3.

Fteimi, N. and Lehner, F. (2018). Analysing and classifying knowledge management publications: A proposed classification scheme. *Journal of Knowledge Management*, *22*(7), Articolo 7. https://doi.org/10.1108/JKM-07-2017-0284.

Gelderman, C.J., Semeijn, J. and Vluggen, R. (2017). Development of sustainability in public sector procurement. *Public Money and Management*, *37*(6), 435–42. https://doi.org/10.1080/09540962.2017.1344027.

Giacomini, D., Rocca, L., Zola, P. and Mazzoleni, M. (2021). Local Governments' environmental disclosure via social networks: Organizational legitimacy and stakeholders' interactions. *Journal of Cleaner Production*, *317*, 128290. https://doi.org/10.1016/j.jclepro.2021.128290.

Greiling, D. and Grüb, B. (2014). Sustainability reporting in Austrian and German local public enterprises. *Journal of Economic Policy Reform*, *17*(3), 209–23. https://doi.org/10.1080/17487870.2014.909315.

Grossi, G. and Argento, D. (2022). The fate of accounting for public governance development. *Accounting, Auditing, and Accountability Journal*, *35*(9), 272–303. https://doi.org/10.1108/AAAJ-11-2020-5001.

Grossi, G., Biancone, P.P., Secinaro, S. and Brescia, V. (2021). Dialogic accounting through popular reporting and digital platforms. *Meditari Accountancy Research*, *29*(7), Articolo 7. https://doi.org/10.1108/MEDAR-01-2021-1163.

Huang, S.K., Kuo, L. and Chou, K.-L. (2018). The impacts of government policies on green utilization diffusion and social benefits: A case study of electric motorcycles in Taiwan. *Energy Policy*, *119*, 473–86. https://doi.org/10.1016/j.enpol.2018.04.061.

Johnson, M.D., Herrmann, A. and Gustafsson, A. (2002). Comparing customer satisfaction across industries and countries. *Journal of Economic Psychology*, *23*(6), 749–69. https://doi.org/10.1016/S0167-4870(02)00137-X.

Khozhylo, I., Lipovska, N., Chernysh, O., Antonova, O., Diegtiar, O. and Dmytriieva, O. (2022). Implementation of smart-city tools as a response to challenges in socio-humanitarian field in Ukrainian metropolises. *Acta Logistica*, *9*(1), 23–30. https://doi.org/10.22306/al.v9i1.262.

Levy, Y. and Ellis, T.J. (2006). A Systems Approach to Conduct an Effective Literature Review in Support of Information Systems Research. *Informing Science: The International Journal of an Emerging Transdiscipline*, *9*, 181–212.

Massaro, M., Dumay, J. and Guthrie, J. (2016). On the shoulders of giants: Undertaking a structured literature review in accounting. *Accounting, Auditing, and Accountability Journal*, *29*, 767–801. https://doi.org/10.1108/AAAJ-01-2015-1939.

Montecalvo, M., Farneti, F. and de Villiers, C. (2018). The potential of integrated reporting to enhance sustainability reporting in the public sector. *Public Money and Management*, *38*(5), 365–74. https://doi.org/10.1080/09540962.2018.1477675.

Montesinos, V. and Brusca, I. (2019). Non-financial reporting in the public sector: Alternatives, trends, and opportunities: La información no financiera en el sector público: Aalternativas, tendencias y oportunidades. *Revista de Contabilidad - Spanish Accounting Review*, *22*(2), Articolo 2. https://doi.org/10.6018/rcsar.383071.

Noyons, E. (2001). Bibliometric mapping of science in a policy context. *Scientometrics*, *50*(1), Articolo 1. https://doi.org/10.1023/A:1005694202977.

Oakleaf, M. (2010). Writing Information Literacy Assessment Plans: A Guide to Best Practice. *Communications in Information Literacy*, *3*(2). https://doi.org/10.15760/comminfolit.2010.3.2.73.

Okoli, C. and Schabram, K. (2010). *A Guide to Conducting a Systematic Literature Review of Information Systems Research. SSRN Scholarly Paper 1954824*. https://doi.org/10.2139/ssrn.1954824.

Petticrew, M. and Roberts, H. (2006). Systematic Reviews in the Social Sciences: A Practical Guide. *Counselling and Psychotherapy Research*, *6*(4), 304–305. https://doi.org/10.1080/14733140600986250.

Secinaro, S. and Calandra, D. (2020). Halal food: Structured literature review and research agenda. *British Food Journal*, *123*(1), Articolo 1. https://doi.org/10.1108/BFJ-03-2020-0234.

Secinaro, S., Calandra, D., Petricean, D. and Chmet, F. (2021). Social Finance and Banking Research as a Driver for Sustainable Development: A Bibliometric Analysis. *Sustainability*, *13*(1), Articolo 1. https://doi.org/10.3390/su13010330.

Secretary-General, U. and Development, World Commission on Environment and Development. (1987). *Report of the World Commission on Environment and Development:* https://digitallibrary.un.org/record/139811.

Střihavková, E., Svobodová, J. and Vysloužilová, D. (2021). Corporate Social Responsibility of Organizations as Part of a Quality Management System. *Production Engineering Archives*, *27*(4), 248–56.

Vara-Sánchez, I., Gallar-Hernández, D., García-García, L., Morán Alonso, N. and Moragues-Faus, A. (2021). The co-production of urban food policies: Exploring the emergence of new governance spaces in three Spanish cities. *Food Policy*, *103*, 102120. https://doi.org/10.1016/j.foodpol.2021.102120.

Walker, H. and Preuss, L. (2008). Fostering sustainability through sourcing from small businesses: Public sector perspectives. *Journal of Cleaner Production*, *16*(15), 1600–1609. https://doi.org/10.1016/j.jclepro.2008.04.014.

Wontner, K.L., Walker, H., Harris, I. and Lynch, J. (2020). Maximising "Community Benefits" in public procurement: Tensions and trade-offs. *International Journal of Operations and Production Management*, *40*(12), 1909–1939. https://doi.org/10.1108/IJOPM-05-2019-0395.

Zioło, M., Filipiak, B.Z., Bąk, I. and Cheba, K. (2022). How Do Environmental, Social, and Governance (ESG) Factors Impact on Public Finance Performance? Risk, Efficiency, and Public Financial System Perspectives. *In*: A. Bem, K. Daszynska-Zygadlo, T. Hajdíková, E. Jáki, and B. Ryszawska (A c. Di), *Sustainable Finance in the Green Economy* (pp. 261–75). Springer International Publishing. https://doi.org/10.1007/978-3-030-81663-6_20.

Zupic, I. and Čater, T. (2015). Bibliometric Methods in Management and Organization. *Organizational Research Methods*, *18*(3), Articolo 3. https://doi.org/10.1177/1094428114562629.

5 Digital Tools for the Public Safety of Future Cities

Mohamed Nour El-Din,[1] Pedro F. Pereira,[2*] João Poças Martins[1] and Nuno M.M. Ramos[2]

1. Introduction

There is an increase in the effect of climate change on the built environment. Cities globally, and cities that are located in low-lying coastal regions especially, are more vulnerable to the effects of climate change due to the concentration of populations in these areas (Riaz et al., 2023), which exceeds the global average by three times (Edmonds et al., 2020). These effects influence extreme weather events, including heatwaves, drought, flash floods, coastal flooding, and sand storms (Ricciardi and Callegari, 2023). Extreme weather conditions (natural disasters) are considered one of the main factors affecting public safety (UN-Habitat, 2012). Thus, improving the public safety of communities is of paramount importance. It has been argued that having a smart disaster/crisis management system is vital for urban safety, supporting decision-making, providing early warning, monitoring, and forecasting to minimize harm from these extreme events (Lacinák and Ristvej, 2017). The indoor environment, particularly the building enclosure, is intended to isolate occupants from the threats caused by the exterior environment, which affects their physical security. However, throughout many of these events and during their subsequent recovery phase, the Indoor Environmental Quality (IEQ) may be compromised, either due to elevated levels of airborne contaminants or difficulties sustaining acceptable indoor environmental conditions (Persily and Emmerich, 2015). Considering that 22% of the population in the EU (95.3 million people) is at risk of poverty or social exclusion, quantifying the effects of extreme weather conditions on energy consumption and health conditions among low-income populations in Europe should be of paramount priority. Owing to the advancements in digital technologies, the concept of future smart cities demonstrates an opportunity to improve public safety, well-being, and quality of life (Rehman et al., 2022).

Smart cities leverage Information and Communication Technologies (ICT) and the Internet of Things (IoTs) to handle extensive data available at the city

[1] CONSTRUCT—GEQUALTEC, Faculty of Engineering (FEUP), University of Porto, 4200-465 Porto, Portugal.

[2] CONSTRUCT—LFC, Faculty of Engineering (FEUP), University of Porto, 4200-465 Porto, Portugal

Email : Mohamed.abushamma@gmail.com; jppm@fe.up.pt; nmmr@fe.up.pt

* Corresponding author: fpfp@fe.up.pt

scale. This data, sourced from various databases, can be utilized by decision-makers and urban practitioners to support various initiatives, such as evaluating the building sector's carbon emissions or simulation of the impacts of extreme weather events on the built environment (Ricciardi and Callegari, 2023). The International Organization for Standardization (ISO) has incorporated a conceptual model of a smart city into its repository (ISO/IEC 30182, 2017). This framework serves as a guide for establishing data interoperability, which is necessary for effective data sharing in urban environments, involving four key types of insight (i.e., operational, critical, analytical, and strategic). However, the data interoperated across the city requires ensuring its availability, quality, compliance, security, and privacy by the decision-makers.

Ideally, cities' modernization is possible by incorporating innovative technologies like Digital Twins (DT), IoTs, Blockchain, and Artificial Intelligence (AI), which are considered vital digital tools in encountering data interoperability issues related to the large amount of data at the city scale. Integrating such tools would allow all stakeholders to acquire accurate, up-to-date information and respond effectively. Rapid advancements in Machine Learning (ML), Swarm Intelligence (SI), Blockchain, and Data Science techniques empowered DTs as a cost-efficient alternative, enhancing the system's overall performance (Jiang et al., 2021). Developing a city Digital Twin will facilitate comprehensive city planning, enhancing disaster resilience (Khallaf et al., 2022). As such, this chapter aims to initiate an enhanced holistic solution for mitigating the impacts of extreme weather conditions on public safety within cities. Establishing a conceptual framework for an Urban Digital Twin (UDT) platform is a foundational step. This platform incorporates both monitoring and controlling IEQ within building enclosures to provide Indoor Environmental Resilience (IER), along with providing an early warning system for extreme weather events. Given that this platform aggregates data from various users, ensuring data security is a priority; thus, a private Blockchain infrastructure is proposed. The chapter is structured as follows: Section 2 explains the state of the art of DTs and UDT in the context of smart cities and climate change. Section 3 introduces the capabilities of SI for the future of UDT. Section 4 proposes a Blockchain-secured UDT framework for mitigating the impacts of extreme weather conditions on public safety within cities. Section 5 includes a description of an early partial implementation of the framework. Finally, Section 6 presents the conclusions and identifies the future trends.

2. Digital Twin and Urban Digital Twin

The concept of DT was initially introduced in the aerospace field in the 1960s when the National Aeronautics and Space Administration (NASA) built at least two simulators to mirror the space conditions to prepare for extensive pre-flight training in the NASA's Apollo program (Rosen et al., 2015). The concept was coined once again almost 20 years ago by Grieves in 2002 as part of a presentation at the University of Michigan for the formation of a Product Lifecycle Management

(PLM) centre; the proposed model consists of three components: real space, virtual space, and linking mechanism for the data/information flow between the real and virtual space; the model was then referred to as "Mirrored Spaces Model" (Grieves, 2014). Afterward, in 2006, the name of the conceptual model proposed by Grieves was changed from the "Mirrored Spaces Model" to the "Information Mirroring Model" (Grieves and Vickers, 2016). Later, in 2010, the first definition for the term DT was introduced by NASA when a DRAFT for an integrated technology roadmap was published under the technology area 11: Modelling, Simulation, Information Technology and Processing, then it was officially published in 2012 stating that "A Digital Twin is an integrated Multiphysics, multiscale, probabilistic simulation of an as-built vehicle or system that uses the best available physical models, sensor updates, fleet history, etc., to mirror the life of its corresponding flying twin" (Shafto et al., 2010, 2012).

Since then, the term DT has been used increasingly in research and industry initiatives. One of the main misconceptions about the definitions of DTs is that some individuals think that a 3D model is a DT. However, to clarify this issue, the study by Kritzinger et al. (2018) highlighted the basis of different levels of integration for DTs, as illustrated in Figure 1.

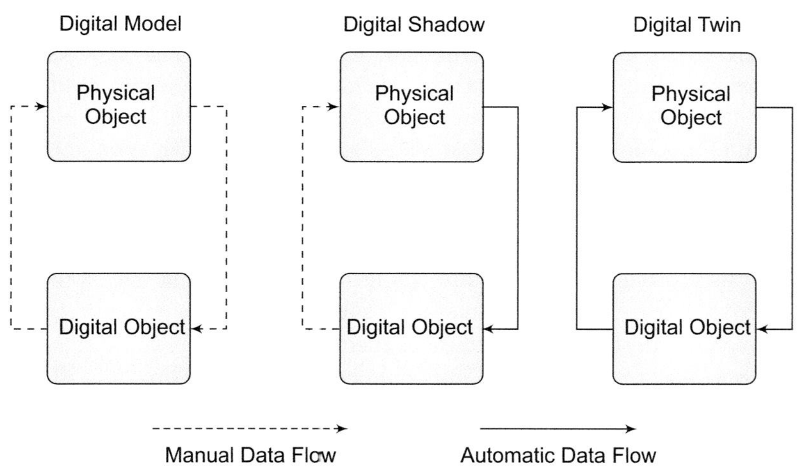

Fig. 1 Differences between Digital Model, Shadow, and Twin based on level of integration (Data flow).

Furthermore, the Centre for Digital Built Britain (CDBB) (Lamb, 2019) differentiated between the two terms Digital Shadow, which was described as a "Static Digital Model" and the Digital Twin as a "Dynamic Digital Model", where data are automatically fed into a dynamic model that can be used for real-time control. In contrast, the data are manually added with time in the static model, which makes it more suitable for strategic planning.

The DT is advocated as one of the supporting technologies that underpin the smart city concept, particularly in the context of urban development. For example, Destination Earth (DestinE) is a major initiative of the European Commission that supports UDT as a part of its aim to achieve a full DT of Earth by virtually representing the city's environment, including buildings, streets, parks, and trees using cutting-edge technologies (European Commission, 2021). Besides, there are developments of UDT in Europe, such as Zurich (Schrotter and Hürzeler, 2020; Stadtmodell, 2022), Rotterdam (Gemeente Rotterdam, 2019), Helsinki (Helsinki, 2019b, 2019a), and Munich (GeoPortal München, 2018). While Rotterdam's UDT lacks direct climate change indicators, it provides energy-saving, green roof, and solar potential data for renewable energy systems, allowing visualization of their impact. Helsinki's UDT, represented by the Energy and Climate Atlas, incorporates various energy and climate-related indicators for each building, considering future climate change projections. Munich's UDT focuses primarily on solar potential but aligns with other cities striving for carbon neutrality and climate resilience. Zurich and Helsinki UDTs integrate climate change parameters into urban transformation simulation, aiding local authorities and decision-makers in planning processes. Though Rotterdam and Munich's UDTs are still in development, they aim to simulate future development scenarios to assess their impact on wind, temperature, solar radiation, and CO_2 emissions. These UDT examples show potential for the future of smart cities. Nevertheless, developing an enhanced UDT that integrates building, neighbourhood, and district models to sensor real-time data to provide early warnings of extreme weather conditions and determine their optimum mitigation and response strategies and an IER in such conditions is not yet acknowledged.

The paradigm of highly efficient sustainable buildings must be achieved with a holistic approach that includes not only the optimization of the building envelope, but also its adequate operation throughout its life cycle. The development of intelligent buildings (IB) will enable lower energy consumption from fossil fuels, increasing the contribution of renewable energy generated *in situ*. In this regard, the Energy Performance of Buildings Directive (EPBD) (Directive 2018/844, 2018), introduced a new building classification index, the SRI (Smart Readiness Indicator), with the aim of improving the use of smart technologies (Verbeke et al., 2020).

The evolution of current buildings towards a paradigm of increasing implementation of intelligent building concepts will necessarily have implications on the design of building envelopes and the selection of systems and equipment. The SRI, highlights a possible organization of the relationship between some areas of building operation (heating, cooling, preparation of domestic hot water, ventilation, lighting, dynamic envelope, electricity, electric vehicle charging, monitoring, and control) and its impacts (energy efficiency, maintenance, comfort, ease of use, health and well-being, occupant information, energy flexibility and storage). This approach, guided by energy efficiency, opens up opportunities of a wide DT application to buildings and to their connection to urban networks and

UDT. Concern for impacts such as comfort and health and well-being require approaches that are not limited to optimizing equipment efficiency. Ventilation control and interaction with the envelope are areas that require a new vision of Building Facilities, which will benefit from this innovation. It should be noted that there are still other areas, such as water efficiency, that must be included in the concerns of the IBs, despite not being directly addressed in the concerns of documents linked to the EPBD. Moreover, using the emerging development of DT for different applications creates the opportunity for a data collection infrastructure that will contribute to public safety digital tools evolution.

3. Swarm Intelligence and the Future of Urban Digital Twins

Digital Twins utilizes data in novel approaches, demanding adaptable IT systems that foster innovation; however, a significant challenge arises from unstructured data that shares a significant portion of big data, posing difficulties for databases designed for structured datasets. Moreover, the escalating risks of security breaches emphasize the need for highly secure and efficient data storage solutions (Margaria and Ryan, 2023). Distributed ledger technologies, such as Blockchain, present a promising alternative, offering decentralized control and enhanced security through distributed storage (Tripathi et al., 2023). This approach addresses security concerns and enables novel scaling techniques like database partitioning and data migration by swarming, fostering a new era of federated and distributed data management conducive to the demands of DTs (Margaria and Ryan, 2023).

The evolution of 'swarm' technology has brought forth a more sophisticated concept known as "Swarm Intelligence", which has garnered extensive attention from researchers since its inception in the 1980s (Ruiz et al., 2023). SI techniques attempt to identify the optimal solution for a given problem by leveraging the collective behaviour of a 'swarm' comprised of homogeneous agents. Although each agent may lack intelligence, the system exhibits self-organizing behaviour, resulting in collective intelligence (Mazzuto et al., 2021).

In the context of UDTs, SI facilitates the development of dynamic and responsive models tailored to the complexities of urban environments. By decentralizing decision-making across a network of agents, each equipped with localized insights and capabilities, UDTs can indistinguishably capture the interactions between various urban elements and environmental variables. The building DTs act as a swarm, where other buildings' DT behaviour influences the behaviour of each building. Alternately, this loosely synchronized interaction helps create semi-orchestrated intelligent behaviour. This approach enables real-time adaptations and optimizations, empowering cities to respond autonomously to changing dynamics such as population flows, energy consumption patterns, and environmental fluctuations. Moreover, SI algorithms provide low computational complexity while surpassing prospects in the accuracy of predictions and computational efficiency (Mazzuto et al., 2021). Thus, by integrating SI into UDTs, cities not only enhance their efficiency and sustainability but also drive innovation in urban design and planning, mirroring the self-organizing principles observed in natural systems.

4. Blockchain-secured Urban DT Framework

Information integration, security, and management on a smart city scale are challenging due to the nature of city complex systems and distributed sources of information besides various interests of the stakeholders (Knezevic et al., 2022). In the context of providing public safety against extreme weather conditions in smart cities, the authors in this section aim to develop a Blockchain-secured UDT framework that utilizes DT, Blockchain, and real-time weather station data to provide early warnings of extreme weather conditions and determine their optimum mitigation and response strategies besides monitoring IEQ conditions to provide insights of the effect of the extreme conditions on the IER. The UDT framework is based on integrating a series of sources (e.g., several DTs of neighbourhoods, private Blockchain infrastructures, and external weather stations) that can communicate and form groups to help make even better decisions by employing the concept of SI. The utilization of SI will help the individual awareness of each DT, coupled with their presence within a shared framework, facilitate the formation of the swarm. While the behaviour of a single DT may seem capricious, the constrained existence in the framework makes the overall behaviour predictable and self-coordinated.

The proposed framework illustrated in Figure 2 consists of four main layers: (1) DT of an Asset; (2) a private Blockchain infrastructure; (3) an external weather station/radar; and (4) a Virtual Data Room. Generally, implementing multiple instances of the first three layers extends the framework from a single DT source to a swarm of DTs forming the base of the UDT.

The framework's foundation lies within the DT layer, serving as the basis for IEQ monitoring and control. This layer focuses on integrating both static and dynamic data. Static data originates from as-built BIM models (IFC files), while dynamic data streams from sensors attached to physical assets, transmitting real-time information to the virtual space within the DT. This approach facilitates rapid identification of IEQ issues, enabling prompt actions and informed decisions to enhance IER through bidirectional dynamic data communication and analytics. Furthermore, real-time data from external weather stations/radars enriches the system, reinforcing the early warning system's effectiveness in mitigating adverse effects on IEQ.

To the authors' knowledge, integrating a private Blockchain infrastructure utilizing BigchainDB software within the Blockchain service layer (BigchainDB GmbH, 2018), DT of assets, and weather stations/radars is a novel move at a city scale.

This integration offers the benefits of both Blockchain and distributed Database technologies, including decentralization, immutability, low latency, and permissioned access. These advantages help address challenges associated with data storage, high transaction costs, confidentiality concerns, and the execution of complex evaluations and optimization that is unfeasible on consortium (public) Blockchains.

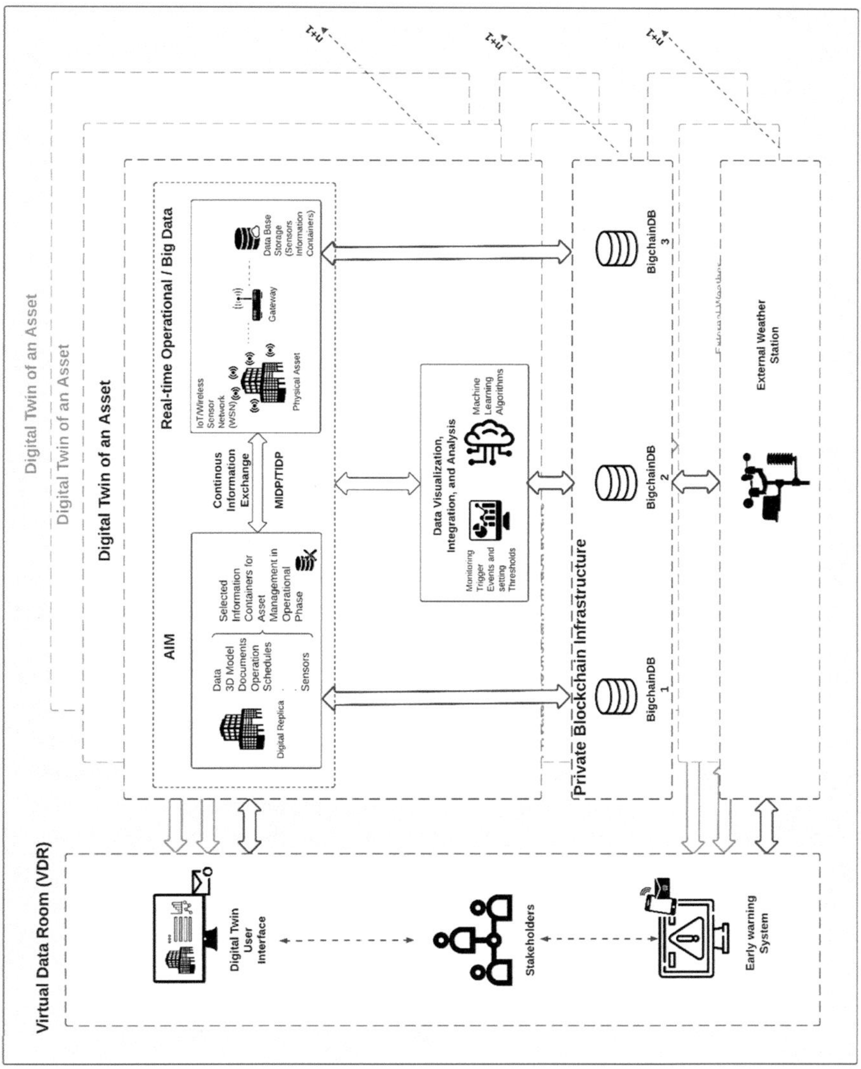

Fig. 2 Blockchain-secured Urban DT Framework.

As part of the framework, the Virtual Data Room furnishes stakeholders with a user-friendly interface to access permissioned information, interact with IEQ data, trigger actions, and receive early warning notifications. Overall, the authors believe that the proposed framework has the ability to effectively manage public safety in future smart cities, facilitating real-time adjustments, flexibility, and independent decision-making.

4.1 Asset DT Layer

The DT layer functions as a unified platform, enabling real-time monitoring, simulation, and analysis to enhance IEQ, predictions, and informed decision-making. Given the challenges stemming from the lack of standardized guidelines for implementing DT, the approach employed in this framework is built upon a pre-existing standardized DT framework developed by the authors (Nour El-Din et al., 2022). This standardized framework incorporates the utilization of BIM ISO 19650 standards within DT processes, enhancing interoperability between digital technologies such as DT and Blockchain. Comprising three sublayers, the first involves physical/built assets, including sensors and actuators, monitoring and controlling various building functions by capturing real-time conditions in order to assess IER. Under extreme weather conditions, several problematic indoor exposures were noted, including dampness, moisture, and flooding; indoor contaminants; thermal stress; infectious agents and pests; and building ventilation, weatherization, and energy use (Persily and Emmerich, 2015). Thus, it is essential to focus on features of the IEQ that comprise four key elements: indoor air quality (IAQ), thermal comfort, acoustics, and illumination (e.g., temperature, relative humidity, light intensity, Total Volatile Organic Compounds (TVOC), particulate matter, and motion) (ASHRAE, 2023). The Asset Information Model (AIM) includes essential information containers, focusing on buildings' energy performance and IEQ management. The information containers include sensor data for real-time energy and IEQ related updates. If the Project Information Model (PIM) of the physical asset exists, unnecessary information is filtered out to reduce stakeholder burdens. The Data Visualization, Integration, and Analysis sublayer offer solutions for processing raw data into actionable insights using data analytics, Artificial Intelligence (AI), and Machine Learning (ML) algorithms. The system integrates a dynamic display feature within the Virtual Data Room, facilitating automated control, feedback adjustment, scenario forecasting, and recommendations.

4.2 Weather Station/Radar Layer

Developing an early warning system is crucial for mitigating the impacts of extreme weather events on occupants, providing timely alerts and accessible information to stakeholders to identify impending risks. However, in many cases, the window for preventative action is limited, highlighting the importance of real-time data as an early warning signal. The foundation of the end-to-end early warning alert structure lies in real-time data collection to predict hazard severity, onset time, and scope (Riaz et al., 2023). Besides, the opportunity to apply DT technology in early warning systems is accompanied by advancements in AI and ML technologies (Kaur et al., 2019).

This study uses weather stations/radars as real-time data sources regarding weather parameters such as rainfall, temperature, humidity, and water vapour pressure. The real-time data is stored in the private Blockchain infrastructure (i.e., BigchainDB). This sensor data is transformed into insightful information with DT's statistical and predictive analysis capabilities. Utilizing an advanced alerting system built on predefined thresholds enables stakeholders to make more timely and informed decisions, as well as providing valuable insights for evaluating the impact of extreme weather conditions on IEQ.

4.3 Blockchain Layer

The Blockchain acts as a layer that guarantees the integrity, transparency, and accountability of data within the system. It effectively governs all transactions within the DT, thereby boosting the trustworthiness of its information. In order to have a reliable early warning system and evaluate the IEQ effectively, accurate and securely exchanged data is paramount. Weather stations/radars and IoTs attached to physical assets are reliable and automated data sources, minimizing human errors. Data is then stored in a central repository known as an "Information container", aligning with BIM processes defined by the ISO 19650 series. However, storing data on public Blockchains poses challenges such as high transaction costs and privacy concerns (Hunhevicz et al., 2022). Thus, prioritizing heightened trust, which may entail higher expenses for storing data on the Blockchain, and choosing off-chain storage, which may result in diminished trust levels, is always a debate that requires a trade-off between trust and storage costs.

The framework introduces a private permissioned Blockchain sublayer between the DT and weather stations/radars layers, leveraging BigchainDB for control and privacy. BigchainDB combines Blockchain advantages (e.g., decentralization, immutability) and distributed database benefits (e.g., low latency, high transaction rate) (BigchainDB GmbH, 2018). IEQ parameters stored in the information containers can be saved to various BigchainDB databases, ensuring confidential information access rights while maintaining data privacy. This approach facilitates reliable and trustworthy data transactions within the DT and guarantees critical updates necessary for prompt decision-making. Furthermore, it enables bi-directional data exchange between algorithmic decisions stored on the Blockchain and their effects on both models and corresponding physical assets.

4.4 VDR Layer

The Virtual Data Room (VDR) is a platform through which stakeholders can access authorized information, interact with performance data and IEQ indicators, view simulation results, and initiate actions via the DT user interface. Moreover, it includes an early warning system based on external weather conditions that are conveyed through real-time data to promptly report to stakeholders (e.g., alert message notifications). This platform fosters collaboration and communication among stakeholders (e.g., disaster management decision-makers, governmental policymakers, and building occupants), enabling them to efficiently utilize pertinent information for their respective roles in mitigating the effects of climate change (i.e., extreme weather events) to their safety and well-being.

5. DT Platform

The Digital Twin developed can be generically described in two ways. Considering the hardware and software, the developed Digital Twin comprises a monitoring system, a communication system, and an interactive platform in all the devices (Figure 3). However, considering the functions performed, the Digital Twin platform developed (https://digitalbuildingtwin.fe.up.pt) currently allows the integration of three essential components in building-level case study analysis: monitoring, data processing and storage, and presentation of the results after the knowledge being extracted from the databases.

The platform features a homepage represented by a map displaying all buildings under study (Figure 4a). Case studies are introduced on the map based on latitude and longitude coordinates. Each case study allows the placement of an individualized IFC file for the floors associated with the installed monitoring system's location (Figure 4b). The platform architecture enables the association of various sensors from different levels. Currently, only dwelling, building, and neighbourhood levels are in use, but there is the possibility of associating sensors related to urban infrastructure to achieve the city level.

The monitoring system in use allows data acquisition for assessing six fields of the building behaviour: thermal comfort, indoor air quality, visual comfort,

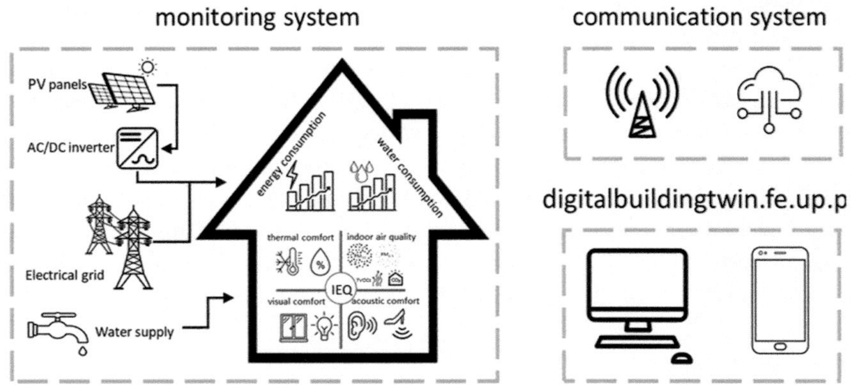

Fig. 3 Architecture of the digitalbuildingtwin.fe.up.pt platform.

acoustic comfort, energy efficiency, and water efficiency (Figure 3). For this purpose, a small box measuring 11.7 cm × 12.5 cm × 8.5 cm was created, housing 10 sensors to measure 13 parameters (Figure 5). This box facilitates the evaluation of the four vertices of Indoor Environmental Quality (IEQ): thermal comfort, indoor air quality, visual comfort, and acoustic comfort. The monitored parameters include the following: temperature, relative humidity, CO_2, particulate matter (PM1.0, PM2.5, and PM10), TVOCs, formaldehyde, occupancy, illuminance in horizontal and vertical planes, sound pressure level, and wind speed. To prevent the conditioning of monitored values, temperature, and relative humidity sensors were placed outside the box, connected by a cable, and internal partitions were created to facilitate air entry for the sensors analysing air quality (CO_2, PMs, TVOCs, and formaldehyde). To monitor energy efficiency, two types of sensors were integrated into the Digital Twin platform: plug devices directly inserted into household sockets in which the plugs of the electrical equipment directly connect, and a totalizer of the dwellings' electrical energy consumption installed directly on each unit meter and, if available, renewable energy production systems. This totalizer can analyse the renewable energy production, the fraction consumed by the dwelling/building/neighbourhood, and the fraction injected into the electrical grid. To evaluate water consumption, water meters were attached to the water supply of each dwelling and placed downstream of each water counter. Monitored data is stored in a Maria DB database and can be accessed, downloaded, or viewed through the platform.

A communication system was created to connect the monitoring system installed in case studies with the Digital Twin platform. The communication system was based on multi-platform single-board mini-computers that send monitored data to a server where it is stored in a MariaDB database. Data transmission is performed via a router linked to the internet using a 4G mobile network modem. The interactive platform with the end user is divided into a backend and a frontend. The backend of the digitalbuildingtwin.fe.up.pt application comprises several services. The main application was implemented in Python, using the Django framework to establish the connection between the API and the MariaDB database. Another Python application was developed to process information from the MQTT service for communication with IoT devices, and a Node.js application was created to convert IFC files. All services and applications run separately in Docker containers. The web application (frontend) is built in Typescript using the React framework, also utilizing libraries such as Xeokit (for BIM navigation), Echarts (for charts), mapbox-gl (for maps), and styled-components/Material UI (for CSS/style) among others. The compilation of the application into HTML/CSS/JavaScript for production is done using Vite.

The platform features a field allowing data processing and the creation of indices based on algorithms whose variables are the monitored parameters. These indices enable the platform adaptation to different monitored buildings and the objectives defined for each case study. The purpose of these indices is to help occupants improve indoor environmental quality conditions and optimize their

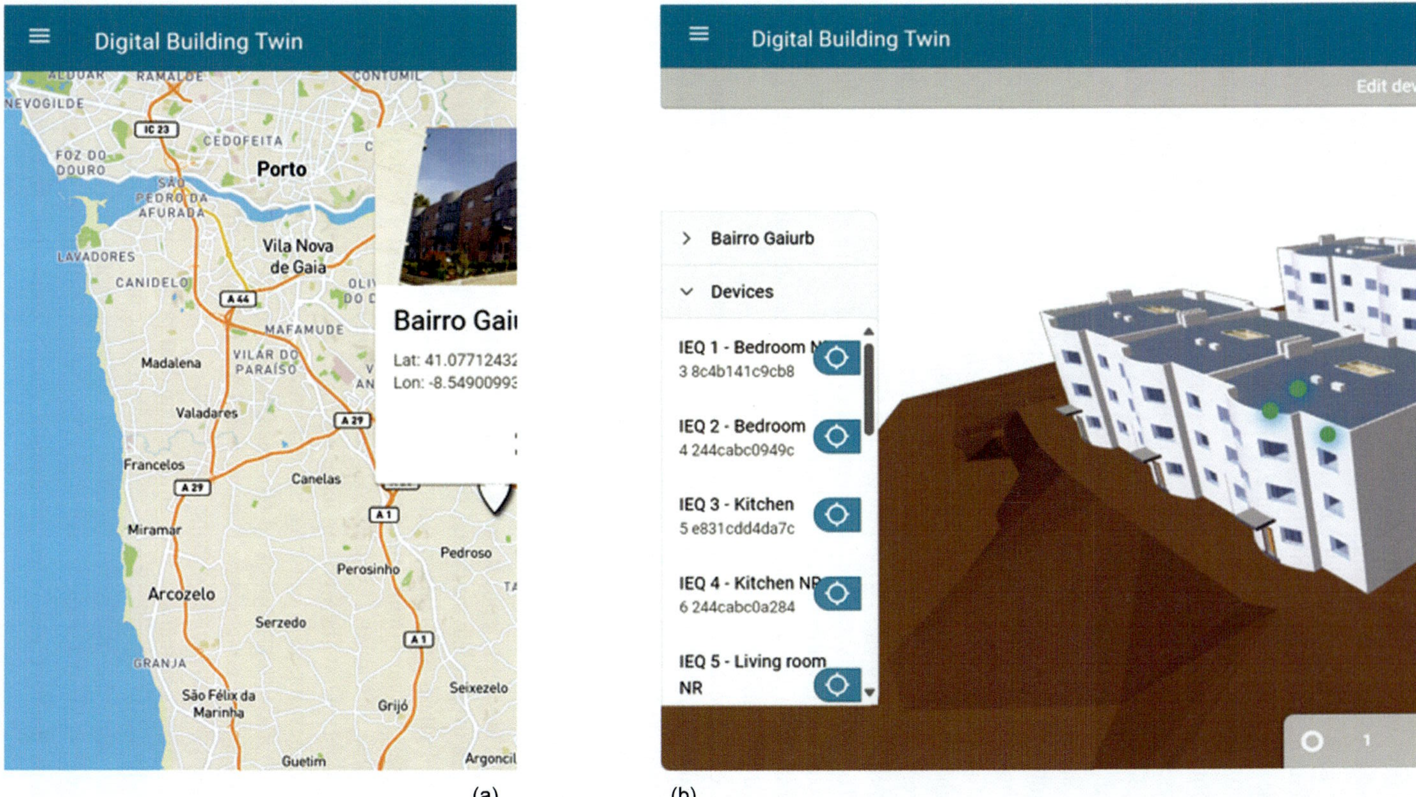

Fig. 4 Initial page of the digitalbuildingtwin.fe.up.pt platform (a), and BIM model with the dots representing the sensors placement (b).

energy and water consumption. This can be done through comparative analyses between fractions with similar exterior envelope conditions, taking into account their occupancy. These analyses allow drawing conclusions about the operation of the dwellings and providing information to occupants to help them improve the dwelling performance. Moreover, in cases of public and/or public-private ownership buildings, it provides building stock managers with information regarding the indoor built environment of each unit and supports data-centric rehabilitation decisions. Building stock managers often have a portfolio of buildings requiring renovation, and establishing prioritization criteria can be challenging. This platform enables an objective analysis based on real monitored data, facilitating decision-making. This information will also be used for post-occupancy evaluations within the scope of smart contracts. This is particularly important for buildings with occupants at high risk of energy poverty, where European funds financially support renovation works. In many cases, these renovation works do not lead to a reduction in energy consumption, as it is already extremely low in terms of heating, but significantly improve indoor environmental quality conditions.

Fig. 5 IEQ monitoring system.

A Digital Twin corresponds to the digital replica of a real model. However, given the functional multiplicity of the built environment and the importance of living beings to represent these real models, achieving complete and cross-cutting coverage of all aspects in a Digital Twin is tough. The digitalbuildingtwin.fe.up.pt platform developed currently enables the creation of a Digital Twin in the fields of IEQ (Indoor Environmental Quality) and energy and water consumption. However, the versatility of the built platform allows for the expansion of these areas to all that the end user needs.

Conclusion

As cities struggle with the increasing challenges posed by climate change and extreme weather events, integrating digital tools emerges as a promising solution for enhancing public safety in future smart cities. A comprehensive framework has been developed to address these pressing issues using UDT and Blockchain technologies. This framework encompasses an early warning system and monitoring of IEQ aimed at strengthening indoor environmental resilience in the face of extreme weather events. Leveraging the DT of assets, a private Blockchain infrastructure, and real-time data from weather stations and radars, stakeholders are equipped with timely and accurate information to implement mitigation measures effectively. Additionally, the VDR serves as a centralized platform, facilitating an enhanced user interface and enabling better utilization of collected data.

Furthermore, an early partial application of a DT platform has been initiated in a neighbourhood in Gaia, Portugal. This initiative involves the installation of indoor sensors to measure IEQ parameters (e.g., temperature, RH, PM, and light intensity), providing valuable data to further enhance the resilience of urban environments to climate change and extreme weather events. By integrating real-time data from these sensors into the Digital Twin framework, stakeholders gain deeper insights into the dynamics of indoor environmental conditions, enabling proactive measures to be implemented swiftly. This localized implementation demonstrates the efficacy of digital tools in enhancing public safety and resilience at the neighbourhood level, laying the groundwork for broader adoption across future smart cities worldwide. Through collaborative efforts and continuous innovation, these digital solutions pave the way for a more sustainable and resilient urban future in the face of environmental challenges.

Acknowledgements

This research was funded by: Base Funding - UIDB/04708/2020 of the CONSTRUCT - Instituto de I&D em Estruturas e Construções - funded by national funds through the FCT/MCTES (PIDDAC), and programmatic funding of the CONSTRUCT, Instituto de I&D em Estruturas e Construções, funded by national funds through the FCT, grant number UI/BD/151302/2021. The second author acknowledges the support of FCT-Fundação para a Ciência e a Tecnologia through the individual Scientific Employment Stimulus 2021.02686.

References

ASHRAE. (2023). *Guideline 10-2023 Interactions Affecting the Achievement of Acceptable Indoor Environments*. American Society of Heating, Refrigerating, and Air-Conditioning Engineers, Inc.

BigchainDB GmbH. (2018). *Whitepaper: BigchainDB 2.0 The Blockchain Database*. https://www.bigchaindb.com/whitepaper/bigchaindb-whitepaper.pdf.

Directive 2018/844. (2018). Directive (EU) 2018/844 of the European Parliament and of the Council of 30 May 2018 amending Directive 2010/31/EU on the energy performance of buildings and Directive 2012/27/EU on energy efficiency. *In*: *Official Journal of European Union*. https://eur-lex.europa.eu/legal-content/EN/TXT/PDF/?uri=CELEX:32018L0844&qid=1666886273533&from=EN.

Edmonds, D.A., Caldwell, R.L., Brondizio, E.S. and Siani, S.M.O. (2020). Coastal flooding will disproportionately impact people on river deltas. *Nature Communications*, *11*(1), 1–8. https://doi.org/10.1038/s41467-020-18531-4.

European Commission. (2021). *Destination Earth*. European Commission. https://digital-strategy.ec.europa.eu/library/destination-earth.

Gemeente Rotterdam. (2019). *Rotterdam 3D*. https://www.3drotterdam.nl/#/

GeoPortal München. (2018). *Energieportal*. https://geoportal.muenchen.de/portal/energie/#/

Grieves, M. (2014). *Digital Twin : Manufacturing Excellence through Virtual Factory Replication*. https://www.researchgate.net/publication/275211047_Digital_Twin_Manufacturing_Excellence_through_Virtual_Factory_Replication.

Grieves, M. and Vickers, J. (2016). Digital twin: Mitigating unpredictable, undesirable emergent behavior in complex systems. *Transdisciplinary Perspectives on Complex Systems: New Findings and Approaches*, *August*, 85–113. https://doi.org/10.1007/978-3-319-38756-7_4.

Helsinki. (2019a). *Helsinki - Energy and Climate Atlas*. https://kartta.hel.fi/3d/atlas/#/

Helsinki. (2019b). *The Kalasatama Digital Twins Project, the Final Report of the KIRA-digi Pilot Project*. https://www.hel.fi/static/liitteet-2019/Kaupunginkanslia/Helsinki3D_Kalasatama_Digital_Twins.pdf.

Hunhevicz, J.J., Motie, M. and Hall, D.M. (2022). Digital building twins and Blockchain for performance-based (smart) contracts. *Automation in Construction*, *133*, 103981. https://doi.org/10.1016/j.autcon.2021.103981.

ISO/IEC 30182. (2017). Smart city concept model: Guidance for establishing a model for data interoperability.

Kaur, M.J., Mishra, V.P. and Maheshwari, P. (2019). The Convergence of Digital Twin, IoT, and Machine Learning: Transforming Data into Action. *In*: *Digital Twin Technologies and Smart Cities*, 3–17. Springer International Publishing. https://doi.org/10.1007/978-3-030-18732-3.

Khallaf, R., Khallaf, L., Anumba, C.J., and Madubuike, O.C. (2022). Review of Digital Twins for Constructed Facilities. *Buildings*, *12*, 2029. https://doi.org/https://doi.org/10.3390/buildings12112029.

Knezevic, M., Donaubauer, A., Moshrefzadeh, M. and Kolbe, T.H. (2022). Managing Urban Digital Twins with an Extended Catalog Service. *ISPRS Annals of the Photogrammetry, Remote Sensing and Spatial Information Sciences*, *10*(4/W3-2022), 119–26. https://doi.org/10.5194/isprs-annals-X-4-W3-2022-119-2022.

Kritzinger, W., Karner, M., Traar, G., Henjes, J. and Sihn, W. (2018). Digital Twin in manufacturing: A categorical literature review and classification. *IFAC-PapersOnLine*, *51*(11), 1016–1022. https://doi.org/10.1016/j.ifacol.2018.08.474.

Lacinák, M. and Ristvej, J. (2017). Smart City, Safety, and Security. *Procedia Engineering*, *192*, 522–27. https://doi.org/10.1016/j.proeng.2017.06.090.

Lamb, K. (2019). *Principle-based Digital Twins: A Scoping Review*. https://www.cdbb.cam.ac.uk/files/scopingreview_dec20.pdf.

Margaria, T. and Ryan, S. (2023). Data and Data Management in the Context of Digital Twins. *In*: N. Crespi, A.T. Drobot, and R. Minerva (Eds.), *The Digital Twin* (Vols. 1–2, 253–78). Springer Nature Switzerland AG. https://doi.org/10.1007/978-3-031-21343-4.

Mazzuto, G., Ciarapica, F.E., Ortenzi, M. and Bevilacqua, M. (2021). The digital twin realization of an ejector for multiphase flows. *Energies*, *14*(17). https://doi.org/10.3390/en14175533.

Nour El-Din, M., Pereira, P.F., Poças Martins, J. and Ramos, N.M.M. (2022). Digital Twins for Construction Assets Using BIM Standard Specifications. *Buildings*, *12*(12). https://doi.org/10.3390/buildings12122155.

Persily, A. and Emmerich, S. (2015). Indoor environmental resilience: A review and discussion. *In*: *Proceedings - Healthy Buildings 2015 America: Innovation in a Time of Energy Uncertainty and Climate Adaptation, HB 2015*, 110–113.

Rehman, A., Shahzad, F., Bin, Y., Razzak, I., Jalil, Z. and Xu, G. (2022). Future smart cities: Requirements, emerging technologies, applications, challenges, and future aspects. *Cities*, *129*(January), 103794. https://doi.org/10.1016/j.cities.2022.103794.

Riaz, K., McAfee, M. and Gharbia, S.S. (2023). Management of Climate Resilience: Exploring the Potential of Digital Twin Technology, 3D City Modelling, and Early Warning Systems. *Sensors*, *23*, 2659. https://doi.org/https://doi.org/10.3390/s23052659.

Ricciardi, G. and Callegari, G. (2023). Digital Twins for Climate-Neutral and Resilient Cities. State of the Art and Future Development as Tools to Support Urban Decision-Making. *In*: E. Arbizzani,

E. Cangelli, C. Clemente, F. Cumo, F. Giofrè, A.M. Giovenale, M. Palme, and S. Paris (Eds.), *Technological Imagination in the Green and Digital Transition*, 617–26. Springer International Publishing.

Rosen, R., Von Wichert, G., Lo, G. and Bettenhausen, K.D. (2015). About the importance of autonomy and digital twins for the future of manufacturing. *IFAC-PapersOnLine*, *28*(3), 567–72. https://doi.org/10.1016/j.ifacol.2015.06.141.

Ruiz, D.E., Gauna, D., Sánchez, L.E., Ruiz-iniesta, A., Villalonga, C. and Serrano, M.A. (2023). Towards an integrated swarm intelligence framework for urban mobility: A systematic review and proposed theoretical model. *Journal of King Saud University - Computer and Information Sciences*, *35*(10), 101836. https://doi.org/10.1016/j.jksuci.2023.101836.

Schrotter, G. and Hürzeler, C. (2020). The Digital Twin of the City of Zurich for Urban Planning. *PFG - Journal of Photogrammetry, Remote Sensing and Geoinformation Science*, *88*(1), 99–112. https://doi.org/10.1007/s41064-020-00092-2.

Shafto, M., Conroy, M., Doyle, R., Glaessgen, E., Kemp, C., LeMoigne, J. and Wang, L. (2010). DRAFT Modeling, Simulation, Information Technology, and Processing Roadmap - Technology Area 11. *In*: *NASA Office of Chief Technologist*. https://www.nasa.gov/pdf/501321main_TA11-MSITP-DRAFT-Nov2010-A1.pdf.

Shafto, M., Conroy, M., Doyle, R., Glaessgen, E., Kemp, C., LeMoigne, J. and Wang, L. (2012). *Modeling, Simulation, Information Technology and Processing Roadmap - Technology Area 11* NASA.

Stadtmodell. (2022). *3D-Stadtmodell. Geomatik + Vermessung Stadt Zürich*. https://www.stadt-zuerich.ch/content/ted/de/index/geoz/plan-und-datenbezug/3d-stadtmodell.html#

Tripathi, G., Abdul, M. and Casalino, G. (2023, March). A comprehensive review of Blockchain technology: Underlying principles and historical background with future challenges. *Decision Analytics Journal*, *9,* 100344. https://doi.org/10.1016/j.dajour.2023.100344.

UN-Habitat. (2012). *Safer Cities Programme*. https://unhabitat.org/programme/safer-cities.

Verbeke, S., Aerts, D., Reynders, G., Ma, Y. and Waide, P. (2020). *Final Report on the Technical Support to the Development of a Smart Readiness Indicator for Buildings - Summary* (Issue June). https://op.europa.eu/en/publication-detail/-/publication/bed75757-fbb4-11ea-b44f-01aa75ed71a1/language-en.

6 Sensors and Networks for Savings and Comfort of Cities' Inhabitants

Bruno Mataloto,[1] João Carlos Ferreira[2] and Ricardo Pontes Resende[3]

1. Introduction

In recent years, the concept of smart cities has gained significant importance as urban centers around the globe seek to harness technological advancements to improve the quality of life for their inhabitants. A key aspect of this shift towards smart urban environments is the integration of sensors, sensor networks, and communication technologies. These technologies play a crucial role in gathering and disseminating real-time data, enabling cities to make informed decisions and provide a comfortable living environment for their residents, (Kummitha and Crutzen, 2017).

Due to population increase and economic progress, energy consumption has been rising over time. The total energy demand worldwide in 2021 was roughly 24,700 TWh (IEA, 2022). Oil was the main source of energy, followed by coal, natural gas, and other fossil fuels. Although they are gaining importance, renewable energy sources like solar and wind power still only account for a small portion of global energy use. In 2021, 28% of the world's energy consumption came from renewable sources (Perea-Moreno, 2021).

The environment and human health may suffer from continuous excessive non-renewable energy use in several ways. Fossil fuel combustion, which accounts for most of the world's energy consumption results in excessive greenhouse gases, like carbon dioxide, being released into the atmosphere. These gases trap heat, raising global temperatures and altering ecosystems, weather patterns, and sea levels as a result (Demin et al., 2014). Air pollution is another problem associated with the use of fossil fuel for energy production, which results in the release of dangerous pollutants such as particulate matter and nitrogen oxides, that can damage the lungs and other organs and lead to several health issues (Reddy and Battu, 2019). The use of foreign non-renewable, resources such as oil, gas, or liquefied natural gas, together with unstable energy prices are also two further economic impacts of using fossil fuels to deal with high energy demands (Ji et al.,

[1,2,3] ISTAR, ISCTE-IUL
Email: Joao.Carlos.Ferreira@iscte-iul.pt; Ricardo.Resende@iscte-iul.pt
* Corresponding author: Bruno_Mataloto@iscte-iul.pt

2019). It is crucial to strive toward reducing energy consumption and boosting the usage of renewable energy sources since excessive energy use is a severe problem that has a major impact on the environment and human health.

The Internet of Things (IoT) is a rapidly growing field that encompasses a wide range of technologies and applications. It refers to the network of physical devices, vehicles, and other objects that are embedded with sensors, software, and other technologies to connect and exchange data over the internet. IoT has the potential to revolutionize various industries, from manufacturing and healthcare to transportation and smart cities.

Systems powered by the IoT have become increasingly important in our cities due to growing concerns about the environmental impact of energy consumption, the finite nature of fossil fuel resources, and the quality-of-life improvement in smart cities. These systems can range from simple devices that allow users to manually control energy use, to more complex systems that use artificial intelligence to optimize energy consumption (Khan et al., 2022), at people's houses, businesses, or even the city's infrastructure.

The importance of IoT systems is based on several reasons. First, they can reduce the amount of energy required to operate buildings and appliances, which can save both homeowners and businesses a lot of capital. Second, energy-saving devices can aid in the fight against global warming by lowering greenhouse gas emissions, and third, these systems have been proven to achieve substantial results regarding people's comfort on multiple fronts (temperature, sound, light, or air quality) and influence human behavior (Mataloto et al., 2023). IoT systems can also increase the resilience and stability of electricity and water grids by analyzing data, preventing waste, detecting leaks, and autonomously produce a response to such events as warnings or even controlling remote devices able to physically interact with the grids.

The development, implementation, and maintenance of IoT systems can not only contribute to the creation of new jobs, and scientific progress in other areas that can benefit from these system results but also promote economic growth, people's self-awareness, and potentially change behaviors and consumption habits. (Nižetić et al., 2019).

This chapter will explore the potential of IoT systems in enhancing comfort within smart cities, such as in public spaces, shared spaces, and households. By examining the role of key elements of these systems, such as sensors, sensor networks, and communication technologies, the chapter will demonstrate how they contribute to creating a more comfortable, efficient, and sustainable urban environment.

2. Comfort Perception

The experience of comfort is a multifaceted phenomenon that encompasses various elements, such as environmental, psychological, and bodily components. Comfort perception is influenced by a variety of physiological reactions, psychological factors, and environmental stimuli, even though it may seem subjective and very personal. Gaining knowledge of these elements might help you better understand how comfort is felt and perceived (Salamone et al., 2018).

From a physiological perspective, the body's capacity to control temperature, humidity, and other environmental variables is intimately related to how comfortable a person feels. For instance, being in extremely hot or cold temperatures or high humidity levels can cause discomfort and negatively affect general health. On the other hand, a cozy setting that keeps the temperature within a consistent, ideal range can improve comfort perception and encourage well-being. Studies have identified a significant correlation of room temperature with thermal comfort as expected, but also with noise perception and overall acoustic comfort. A significant effect was also found for the interactions of both room temperature and sound levels with air quality perception, highlighting the importance of air quality perceptiveness in achieving occupants' comfort and satisfaction with office space (Bourikas et al., 2021).

The impression of comfort is also significantly influenced by psychological variables. The way that people feel comfortable can be influenced by their cultural background, personal preferences, and past experiences. A person from a colder climate might enjoy colder surroundings, whereas someone from a warmer region might feel more at ease in hotter temps. Furthermore, individual preferences for music, lighting, and other aspects of the surroundings might affect how comfortable someone feels. In Castaldo et al. (2018), 80% of the enquired employees, were declared to be positively influenced by the pleasant aesthetics and liveability of the workplace.

The way a room is designed, the standard of the furniture, and the presence of facilities are examples of external factors that might affect how relaxed and satisfied a person feels. A poorly planned place with uncomfortable sitting and harsh lighting, on the other hand, can reduce comfort levels. As an example, a well-designed office with comfortable furniture, sufficient lighting, and optimal thermal conditions, not only promotes comfort but also work efficiency and general performance (Hoque and Weil, 2016). The fact that each person has their optimal comfort level for determined environmental conditions indicates that inside a room, workspace locations can have a direct impact on their performance and that same impact can be minimized if evaluated properly from the room perspective (thermal/light and sound conditions of each workspace) and from the user perspective (conditions for which the person feels more comfortable with).

3. IOT Systems Architecture

At the core of an IoT system, functionality lies in its architecture, which assures the interaction between the system's components, guaranteeing a stable and reliable operation. A well-designed architecture is crucial for ensuring the scalability, security, and efficiency of an IoT system that typically consists of three main components: the physical layer, the communication layer, and the application layer (Alshohoumi et al., 2019).

The physical layer consists of the physical devices, sensors, and actuators that collect data from the environment. These devices can range from simple sensors that measure temperature or humidity to complex devices such as industrial robots or medical devices.

The communication layer is responsible for transmitting data between the physical devices and the application layer. This layer includes protocols such as Wi-Fi, Bluetooth, Zigbee, and LoRaWAN, which enable devices to communicate with each other and with external systems.

The application layer is where the data collected by the physical devices is processed and analyzed. This layer includes software applications, analytics tools, and cloud-based platforms that process the data and provide insights and recommendations.

When developing an IoT system, some considerations must be taken into account:

1. **Scalability:** Essential for accommodating the growing number of devices and data within an IoT system. The architecture should be designed to expand effortlessly, ensuring uninterrupted functionality as the system evolves.
2. **Security:** Extremely important, especially in sensitive environments like main city grids (energy, water, and gas), healthcare, or industrial applications, robust security measures must safeguard the system from cyber threats, protecting it against data breaches and unauthorized access.
3. **Interoperability:** Enables devices from different manufacturers to communicate. It is imperative to promote collaboration among various systems and devices, ensuring interoperability across the ecosystem to allow multiple IoT systems in a smart city to communicate and exchange data.
4. **Low latency:** Crucial for real-time data processing, forcing minimal delays in data transmission. Edge computing, which is a concept where data is processed closer to its source, and high-speed communication protocols are instrumental in achieving low latency.
5. **Energy efficiency:** Vital for extending the battery life of IoT devices that are required to be deployed in remote areas. Designs that optimize power consumption are essential for ensuring sustained device performance and reducing the frequency of recharging.

4. Different Architectural Approaches

Several architectural approaches can be adopted in IoT systems, each with its advantages and limitations.

Centralized Architecture: A central server that gathers, processes, and saves data from every device connected to the network is what defines an IoT system's centralized design. This design is easy to adopt since it is very simple to implement.

It does, however, have a few drawbacks. Initially, it might not be adequately scalable to manage a large volume of data and devices. Second, given that the entire system could be affected by a central server failure, it might raise reliability concerns. Finally, since data must travel to and from the central server, it can have higher latency, which could cause issues for real-time applications. A centralized architecture is nevertheless commonly used despite these drawbacks, particularly in applications where scalability is less critical than simplicity and convenience of implementation.

Distributed Architecture: In a distributed architecture, instead of being concentrated in a single server or location, data is processed and stored among numerous devices and nodes. Distributing the workload among several devices can enhance scalability and, since a single device failure does not compromise the entire system, it can boost reliability. Because data can be processed locally on nodes or devices that are closer to the source, it can have lower latency. However, because it requires cooperation between nodes and devices, and careful consideration of factors such as data consistency and fault tolerance, a distributed architecture can be more difficult to implement and maintain. Furthermore, it may be harder to safeguard because there are more possible sources of vulnerability. A distributed architecture may be a good fit for applications that need high levels of performance, reliability, and scalability even in the face of these difficulties.

Edge Computing Architecture: Edge computing architecture is a concept that moves data processing and analysis away from a centralized cloud infrastructure and toward the data source, such as IoT devices, sensors, or gateways/access points. Enhanced privacy and security, better bandwidth economy, and decreased latency are some of the advantages of this strategy. Edge computing is perfect for real-time applications because it can drastically cut down on the time it takes for data to travel from the source to the cloud and back by processing it locally, at the 'edge' of the network. Moreover, edge computing can maximize network capacity by aggregating and filtering data locally, forwarding only relevant data to the cloud for additional processing. Furthermore, by lowering the chance of data breaches during data transmission, privacy and security can be increased. Although this architecture can greatly improve the efficiency, reliability, and security of IoT systems, it has significant constraints regarding processing and storage resources.

Fog Computing Architecture: An expansion of edge computing, Fog computing architecture adds fog nodes to the edge devices' computational capabilities. Compared to regular cloud servers, fog nodes are situated farther away from edge devices but closer to them, nonetheless. IoT systems benefit from this architecture in several ways, including decreased latency, increased scalability, and improved reliability. Fog computing is perfect for real-time applications because it can drastically cut down on the amount of time it takes for data to travel from the source to the cloud and back by processing it closer to the edge. Additionally, by dispersing computation among a network of fog nodes, fog computing can increase scalability and provide more effective data processing and analysis than edge-computing alone.

5. Users as Part of the Architecture

The effectiveness of IoT systems often depends significantly on users' ability to interact with the system, understand the information provided, and take appropriate actions, especially when the system's capability to interact with the environment through actuators is limited. Through a variety of interfaces, including physical devices, web portals, and mobile apps, they can communicate with the system to operate and monitor linked devices, access data and analytics, and get alerts and notifications. To provide a good user experience and make it simple for users to access and use the system's functions, user interface design is crucial. To guarantee that only authorized users can access sensitive data or operate devices, user authentication, and authorization procedures are also an essential part of the design of IoT systems. Some techniques used to confirm user identities and authorize access include biometrics, multifactor authentication, and role-based access control.

Through their interactions with the system, users offer the IoT system important information and feedback as well. This input can help find problems with the system, make it work better, and guide future improvements.

6. User Interaction and Behaviour Change

IoT systems that are user-friendly and intuitive are more likely to be used regularly by users. This can be achieved through clear and concise user interfaces, as well as by visual aids such as graphs and charts to help users understand the displayed information and how it can affect the user directly, for example when controlling room temperatures, or indirectly, for example, giving costs predictions (Herrmann et al., 2021).

Users may be more likely to interact with energy management systems if they feel that they will benefit from doing so. For example, users might be more inclined to use the system if it helps them save money on their energy bills or if it allows them to earn rewards or discounts (Soomro et al., 2021). Providing users with feedback and support can also help to increase engagement, for instance, by triggering alerts or notifications when energy consumption exceeds certain thresholds, or the ability to receive support from customer service representatives if users have questions or encounter issues (Böckle et al., 2020).

IoT systems that are prominently displayed or easily accessible are more likely to be used. This might involve placing the system in a central location in the home or office or making it available through a mobile app that can be accessed from anywhere (Inyim et al., 2018).

Educating users about the benefits of energy management systems can help to increase their engagement with the system. This might involve highlighting the environmental benefits of energy conservation or emphasizing the potential financial savings that can be achieved through more efficient energy use (Mi et al., 2020).

In the context of IoT systems, user interaction and interface design can promote beneficial behavioral change. It's critical to examine the establishment of user interaction, the cultural and economic drivers of energy and water savings, the technological advancements supporting user engagement, and the user-centered design concepts that mold user interfaces in order to fully comprehend the complex nature of user interaction.

7. Societal Motivations for Savings

In Portugal, a study (Abreu et al., 2020) highlighted how age groups influence residential savings adoption. Younger homeowners tend to slowly engage in sustainable projects and techniques, doing small renovations at a time. They also represent the demographic group most concerned with environmental efficiency. Older homeowners, despite having greater financial standing, dislike frequent changes to their homes and are less prone to renovate or engage in sustainable projects. Portugal, with a predominantly older homeowner population, should focus public policy on younger demographic groups, who are also more prone to technological solutions, but also on the development of interactive and simple solutions, such as IoT systems, which can also be of use to the older population if designed with all ages and technological literacy levels in consideration.

In Berardi (2013), authors evaluated the construction sector stakeholders' influence on the adoption of IoT technologies in Italian homes and concluded that, despite having the power to promote sustainable construction methods, materials, and technologies, homeowners don't do it. From case studies, some of the reasons identified are lack of information, lack of government incentive policies, or economic reasons.

Li et al. (2023) studied the effects of social relationships on saving behaviors at home, the workplace, and the community. The study uses questionnaire data from over 1000 participants to analyze the role of social interactions, and the level of difficulty of saving behaviors (such as habit adjustments, equipment upgrades, and housing retrofits). The results indicate that social interactions have a positive impact on saving behaviors, but the effect varies depending on the behavior and the level of difficulty. Low-difficulty behaviors are more susceptible to passive interpersonal interactions (when the saving behavior is introduced by someone else), while high-difficulty behaviors are influenced more by public relations and active interactions (when saving behaviors are induced to other people). The study also found that individual factors such as perceived value and information processing depth play a role in the impact of social interactions on saving behaviors. A study about psychological factors that influence energy-related behaviors in office environments (Weerasinghe et al., 2023) was conducted in New Zealand, with 294 questionary responses. The authors analyzed the relationship between motivation, opportunity, and ability energy-related behaviors. The results show that improving opportunities for savings through norms, organizational support, behavioral interventions, and accessibility to control increases occupants perceived behavioral control and knowledge, leading to a stronger attitude and sense of obligation to save resources without compromising comfort. Another study by Du and Pan (2021) described the impact of behavioral interventions on energy use in buildings, specifically cooling-related behaviors in subtropical areas. The authors developed a framework that simulates models of building energy, occupant behavior, and social psychology. The results of the simulation showed that behavioral interventions can lead to energy savings of 15-18% in air conditioning electricity use. However, the effectiveness of the interventions is determined by multiple factors, such as the original behavior, comfort demand, saving beliefs, and the intervention time.

Caferra et al. (2021) highlight the importance of taking into account behavioral aspects when creating energy policies. It argues that trust in institutions and peers can improve social behavior, including responsible energy consumption. The study used data from the 8th round of the European Social Survey to measure political trust, social trust, and saving habits. The results showed that social and political trust has a positive effect on people's willingness to reduce their energy consumption at home and to reduce their negative environmental impact.

8. Hardware and Software-based Solutions

Hardware and software are two common key aspects of all approaches. Many European countries focus their energy efficiency strategies on expanding the use of cleaner energy sources for building and manufacturing industries; however, as Simanaviciene et al. (2014) indicate, household strategies and sustainable domestic energy savings, and sustainable energy use have a clear potential to increase overall energy savings substantially. Results show that the less expensive and more efficient way to reduce greenhouse gas emissions in a household is to reduce energy consumption. While domestic appliances' energy efficiency labelling and building renovations are currently the most effective, authors conclude that many other concrete measures can be taken, such as technological approaches with sensorization and automation.

In the project reported in LaMarche et al. (2014), the authors studied why most energy monitoring systems lose effectiveness over time, even when savings can reach up to 13%. To understand the cause, the authors proposed a flexible and simple system to provide information to users without losing efficiency. The first study of this project concluded that users want to access their system across multiple platforms, both at home and remotely, however, the availability of an application to consult domestic energy consumption does not guarantee that users will be engaged over longer periods. The second study concluded users also prefer content that is easy to understand, unlike charts or text that require time and effort for interpretation. Each user might prefer different types of interfaces or even interact more with some functionalities than others, therefore a customizable system that can be designed according to user preferences can achieve better interaction results. Economic reasons are also why many users are not willing to have a monitoring system, although the price of such technologies has been getting lower each year (Inoue et al., 2022). The third study determined multimedia feedback as the easiest and most pleasing way to display information on an energy monitoring system dashboard, as well as proactive alerts. Researchers also claim that household energy savings depend mainly on education/user behavior and heat consumption efficiency (Ruokamo, 2016), so it is essential to find new strategies to tackle these problems. Most residential energy-saving approaches tend to produce changes using only autonomous mechanisms, sensors, or other smart technologies, with little or no user interaction beyond visualization. According to a systematic review (McIlvennie et al., 2020), user-centric techniques should be created in conjunction with autonomous systems to get an ideal solution, allowing users to participate in the process and subsequently alter their behavior.

Visconti et al. (2019) proposed an energy monitoring system, using sensors, actuators, and a mobile application, where the user can activate wireless plugs and know energy consumption in real time. Sensors were also employed to detect motion and apply rules to turn off actuators based on human presence, but because the system was not designed to show the users what they were gaining from such actions, it required a technician to set up each new rule, and quickly the system became obsolete.

Yang et al. (2022) aimed to reduce the energy consumption of a central air conditioning system in a commercial building in Shenzhen, China. The study used historical operation data and pattern recognition algorithms to identify saving strategies. The results showed that the saving strategies based on decision tree algorithms could conserve more than 30% of the total energy consumption of the building's central air conditioning system. This demonstrates the potential for using data analysis and machine learning algorithms to improve energy efficiency in central air conditioning systems; however, these systems require large amounts of historical data for training, and results may vary according to data quality and detail. Another study performed by Jia Qu (2022) in a university computer room focused on improving the overall service quality and reducing the energy consumption of the equipment. It proposes a solution using IoT to remotely monitor the computer room's thermal environment and energy consumption, although there is no mention of a user interface for sensor monitoring or interaction with technicians remotely.

9. User-centric Approaches

The household transition to low and zero-carbon technologies has reportedly been relatively slow due to misinformation, functionality issues, ergonomic issues, connectivity to other systems, as well as cost and payback time, according to a 2007 survey in the UK (Roy and Caird, 2007) with 400 participants and 90 in-depth telephone interviews. This study highlights the necessity of user feedback in the creation of new systems intended to conserve energy and cut carbon emissions because these systems will be much less effective if users cannot be engaged.

Isnen et al. (2020) present a load optimization system that adapts energy consumption and comfort according to a fixed daily energy budget calculated from 30 days of historical data and user presence. By using low-cost motion and current sensors, the system monitors appliances, lights, and other devices, and if daily maximum consumption is reached, turns them off, unless the motion sensors detect user presence. The proposed algorithm saved 2.62% of energy, however, this value only applies if there is no motion at the house, meaning that users' behaviors will not change or have any impact on energy savings. The system design also wrongfully implies that it is possible to establish a realistic budget every day, not having into account possible disruptions in energy consumption such as extreme weather or more people at the house.

Liu (2013) developed a straightforward method to provide persuading images in accordance with home energy usage; for example, a dazzling flower when consumption is low (positive) and a wilted flower when utilizing is excessive (bad). After a two-week baseline definition period, this method was evaluated for four weeks in 10 different homes. Results revealed that over four weeks, energy savings tended to increase by up to 13%. This study shows how visual feedback may affect users, but it is still essential to provide enough information for users to understand the situation and advise them to act. If the interface is static or barely presents changes over time, users tend to lose interest and abandon the system or application.

García et al. (2017) implemented a framework with both technology and social computing to influence users. The system collects users' locations from Zigbee wireless sensors (Ramya et al., 2011) that monitor energy consumption, temperature, and light in every room. The framework presented the user with suggestions for heating, light state warnings, and other sustainable actions. The system was tested in five apartments for 30 days for the energy baseline and the other 30 days for the experiment. Results show an average of 17% energy saving, with the most effective suggestions regarding heating and lighting. The negative impact of such a system is the implementation cost, which can be a decisive factor for many people, however, each application hardware can be selected based on the requirements and precision required for each environment, lowering implementation costs. Having an average energy savings of 17%, the economic return of such an investment can also be used as motivation for users.

User motivation and behavior are also essential. Onuma et al. (2020) developed a recommendation system based on user activities by understanding their habits through an energy consumption pattern analysis for multiple devices. The system was able to reach a 35% prediction effectiveness and recommend personalized sustainable actions that reduced energy consumption in their household. In this system there was not a clear interaction mechanism with the users, making the predictions irrelevant. The impossibility of remotely controlling devices also sets a limitation on the system since it relies solely on user decisions.

Researchers from Switzerland interacted with users of an app-based energy-saving system for three months, which successfully led to energy savings, behavior changes, and increased user awareness (Wemyss et al., 2019). However, after one year, even though users believed their behavior had changed because of the system's use, energy savings had returned to levels prior to the intervention. In this research, the time factor made the system obsolete and unattractive, resulting in the reversion of energy-saving habits that were developed when using the application. Alam et al. (2016) performed a survey for existing computational methods for energy management in a household environment, considering energy predictions for consumption and production, and autonomous load scheduling based on the predictions.

These complex technological approaches limit the interaction of users with the system. Most systems that require specific expertise can only be maintenance by specialized technicians and this can affect a user's willingness to use the system. The amount of hardware and control over appliances can also influence a user's decision when choosing an energy-saving system.

Despite being able to predict energy consumption/demand/production, based on user behavior and household consumption profiles and control some appliances, these systems lack user-direct interaction and contribution since if the system does not meet the household members' needs/comfort levels, no lasting savings will be achieved, and usage will decline over time.

10. 3D Dashboard Interfaces

Visual feedback has a major impact on energy-saving systems and studies that use a mobile application or visual interaction with users, have better engagement results. 3D Visualization has been developed for many purposes, such as architecture with 3D models, entertainment with game engines, education, medicine, and others like Virtual Reality and Augmented Reality (Boga et al., 2018). With 3D modeling software, it is possible to create simplified models or simulations and integrate sensor information that can turn a simple 3D house model into a dynamic environment with shapes and colors that can shift according to real-time sensor data.

Shahinmoghadam et al. (2021) developed a virtual reality application for thermal comfort assessment using BIM models and real-time simulated sensor data, where users can simulate comfort level changing parameters such as clothing level and metabolic rate or ambient temperature. This approach increases users' perception of how many factors can interfere with a person's thermal comfort, sometimes leading to energy waste by using air conditioning when it is not required. In this study, there is no interaction beyond the simulation, therefore, the user cannot experience real changes in the environment and there are no recommendations or suggestions that could help users to improve their comfort or reduce energy consumption for that purpose.

Systems whose purpose is to reduce energy or resource consumption either rely on predefined rules to control appliances, or human interaction to access and intervene. The problem with this approach, as proven in "Keeping energy visible? Exploring how householders interact with feedback from smart energy monitors in the longer term," by Hargreaves et al. (2013), is that human interaction reduces over time, as user interest wanes, the system loses its novelty and is blended into the daily routine.

Dashboard visual feedback is one of the best ways to help people interact with systems that generate large amounts of data, that would be hard to comprehend without it (Yera et al., 2018). An interactive dashboard can filter, summarize, and present information relevant to the user into simple widgets and charts. Furthermore, Bera (2016) studied how colors in dashboards containing charts and indicators affect users' decision-making (in this case from a business perspective), and results show that color usage attracts viewers' attention; however, this can have a negative impact if the purpose is not to empathize that information. If correctly applied, dynamic-colored dashboards can provide extra meaning to certain data, and draw user's attention, sustaining engagement over time.

Darejeh and Singh (2013) examined approaches to enhance user interfaces for those with lower computer literacy, seniors, kids, and other compromising variables including memorization difficulties. Using graphical elements like avatars turned out to be quite alluring, especially for kids, and appealing in general for regular users. Other recommendations focused on uncluttering UIs (user interfaces) and removing extraneous features and buttons that can make users anxious or hesitant, and frustrated.

Mobile applications exist for almost any purpose that can be thought of. User engagement with mobile apps is an emerging trend since smartphones became personal objects that are always within reach. Mobile applications proved to have a functional, social, and emotional influence on users, using gamification technics (Wut et al., 2021), notifications, and reports. These applications can provide a close connection between an energy-saving system and its users and be used for notifications of events that require user intervention.

11. Case Study in a Smart Home

IoT devices can make homes and buildings more comfortable for users in many ways, such as monitoring environmental parameters using sensors and controlling them using actuators. Sensors and data analytics are used by smart home appliances like lighting controls, security cameras, and thermostats to maximize energy efficiency, boost safety, and increase user comfort. For instance, a Smart Buildings project for a company building called "The Edge" in Amsterdam, Netherlands (Apanaviciene et al., 2020), makes use of IoT sensors to monitor temperature, lighting, energy usage, and occupancy in all of the building rooms. The lighting, heating, and cooling systems are then modified based on sensor data and user input data to achieve the optimal efficiency and comfort for employees.

An IoT system was installed in a Portuguese home and monitored energy consumption and interactions with the system's dashboard over a 12-month period. The house was inhabited by four adults, ages 23-50, who possessed a mediocre level of technological expertise to carry out basic system functions including navigating a general dashboard. Since energy management had always been a priority at the house, the proposed solution could reduce energy use by altering user behavior rather than merely adding new equipment and modifying the settings of already-existing systems, such as the lighting, heating, and cooling.

The Lisbon, Portugal-based residence was constructed in 2001, with all its windows facing north and south. The owners had a photovoltaic microgeneration system with two 330-watt solar panels installed prior to the system's deployment. In addition, the house featured an energy metering system that collected historical data for more than five years prior to the system's implementation. Sensors were used to monitor production and consumption, connect to the Internet, and communicate with the proposed solution. Each room was equipped with a temperature and humidity sensor, a light sensor, and a motion sensor, together with Wi-Fi plugs to control electrical heaters. Sensors were placed 1.5 meters from the ground to reflect the real comfort temperature felt by the users and not the temperature closer to the ceiling measured by air conditioning units, which is higher because heat raises by convection.

The system was composed of a tablet display that was positioned in a visible and conspicuous area, like an entryway or corridor. The tablet displayed a straightforward dashboard (Figure 1) with real-time temperature and humidity readings from each of the house's main rooms as well as real-time energy production and consumption, each with a color code for high, normal, and low values.

Along with daily statistics of energy data, general information such as date, time, and local weather also gave the user an additional reason to glance at the screen.

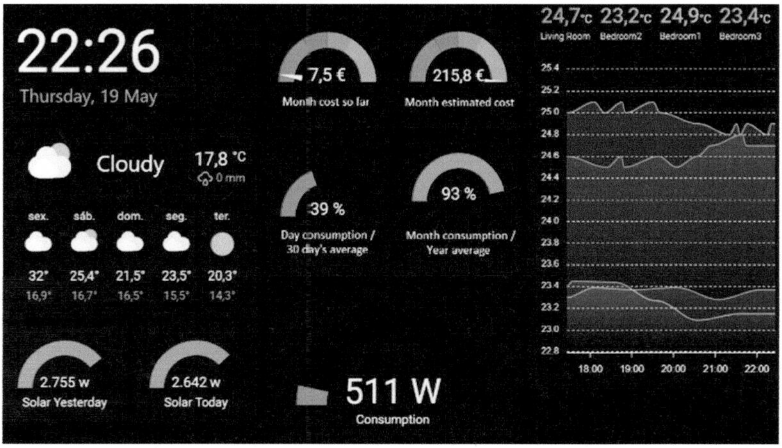

Fig. 1 Main informative dashboard view of the IoT system.

Novelty is a key component of an informative panel that dwindles over time. To address this issue, a utility view was introduced (Figure 2) allowing users to remotely control devices like smart plugs and relay-connected lights, electric heaters, and shutters.

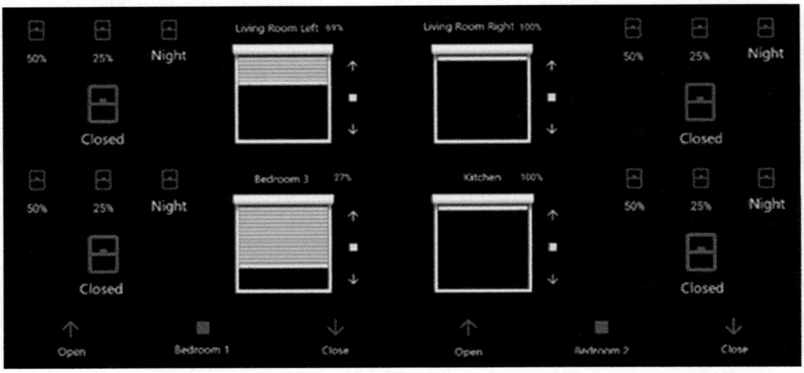

Fig. 2 Utility view for device control.

Because too much intricacy in one view can make it less visually appealing, more views or tabs were added to the dashboard that rotate at different times of the day. Not only it is possible to show more data without overwhelming a single view, but this rolling process can also be used to provide views that are more pertinent at different times of the day. For example, users typically lower the shutters during lunch. As a result, this view can be automatically displayed more frequently during lunch than at other times.

Because heating and cooling appliances represent a major part of the energy bill (Geng et al., 2018), a predefined temperature-control system was developed.

Since most domestic low-cost electric heaters rely on timers to operate, essentially turning the heater on and off for predefined periods of time, they lack a dependable method for controlling the room temperature. The temperature of a room can be more effectively controlled by air conditioners, but over time, their sensors may become less dependable. (Mataloto et al., 2019).

A tailored and controlled room environment can increase comfort and reduce energy usage.

The specified temperature management system precisely and accurately regulates the room temperature by using smart plugs and distributed sensors. At first, the system used the predefined temperatures to define the lowest and maximum temperatures in each room according to existing standards (ASHRAE, 2013). Then, with a smartphone application or home dashboard, any user can modify these variables to suit their needs and preferences.

The system has access to electric shutters, which it uses to close the shutters in rooms facing the sun when the outside temperature rises above a predetermined high threshold if no one is home. This prevents the side of the house from overheating and saves energy when cooling during summer season, and the opposite situation during winter season. Using the contained temperature range for space heating has practical energy consequences that result in a 70% energy savings. The room temperature remains constant with minor fluctuations of less than 0.5°C, preventing overheating, and allowing the user to feel a comfort level customized to his needs.

There were also customized dashboards available that provide a color overlay over a 3D representation of the house. Through a smartphone application (Figure 3), any member of the household could access the 3D dashboard and choose which rooms they wished to receive notifications for in the event of an event, such as lights left on without movement for a predetermined period. Smart plugs, heater settings, and other user-related devices are also shown, enabling intelligent remote control of the room's environment.

Artificial lighting is another simple automated method that can save a substantial amount of energy. By using each room's motion sensor, it is possible to determine whether energy is wasted in an empty room and issue a warning to the user's mobile application.

Fig. 3 3D Interactive Android application for Smart Home.

The developed IoT system focused on the long-term energy consumption, energy production, and household thermal comfort. When users gain awareness of how much energy a certain piece of equipment is used, they begin to question its use and efficiency. This approach combined multiple user interaction techniques to increase awareness, and interaction within UI systems, and enhance long-term engagement. A colored indicator dashboard with which users can interact helps to prevent loss of interest over time, whereas a mobile application with colored avatars and customized notifications keeps the user aware when not at home or in direct contact with the system. To track shifts in energy consumption patterns, we analyzed three years of historical consumption data alongside the year of the system's implementation. This involved conducting a detailed examination of the household's energy profile to identify which appliances consumed the most energy and to pinpoint any deviations from previous years, thereby ensuring accurate data comparison without interference. As a result of this analysis, the average energy consumption was reduced by 19%.

Since the test case was conducted during the COVID-19 pandemic, we were faced with a major problem when finding possible candidates to test the system. Given the complexity of the installation, monitoring, data collection and analysis, we decided to present the test case system on a large-scale survey to collect more statistically relevant data without contacting directly with potential users.

With a total of 215 responses, seven were removed for not being completed, therefore 208 responses were selected for the survey analysis. The survey respondents were primarily males (76.4%), 14% of them were technology experts, the majority had moderate (35%) to advanced technological knowledge (38%), and 13% had basic knowledge. In terms of energy-saving habits, 45% of the respondents reported frequent to very frequent energy-saving habits, and 39% of respondents reported moderate habits. Only 16% claimed minimum to no saving habits at all.

Participants were presented with three energy management system images to assess their attractiveness and utility.

The first system showcased an existing energy monitoring solution available in the market. It featured a real-time energy gauge meter, an estimated energy cost for the current month, and a daily chart illustrating energy consumption.

The second system represented the initial stage of our approach (Figure 3). It displayed real-time energy consumption and production data, along with basic information such as the date, time, and local weather.

The third system depicted the latest version of our proposed approach (Figures 1 and 2). It presented the main dashboard featuring energy production/consumption metrics, statistics on monthly energy costs, indicators for temperature/humidity in each room, and a utility view with controls for electric shutters and smart plugs.

Participants rated each of the three systems for attractiveness and usefulness on a scale from 1–10. The market solution example received the most varied responses, with 54% of participants rating it between 6–8. The second system was perceived as slightly less attractive than the third system, with 54% of ratings falling between 8–10, while the third system garnered the highest attractiveness ratings, with 69% falling between 8–10. Regarding usefulness, most participants identified the third system as the most beneficial, with 78% of responses rating it between 8–10, likely due to its utility-focused approach.

Participants were asked to rate their agreement on a scale from 1 (totally disagree) –10 (totally agree) regarding three aspects of energy management systems: the relevance of data, the utility approach's impact on sustained interest, and the perceived effect on sustainable behaviors. Regarding the relevance of data visualization, a significant majority (80%) rated between 8–10, indicating strong or total agreement. Similarly, when asked if real-time data visualization, including energy usage, temperatures, and other statistics presented on the system, could influence behavior changes, the majority (81%) strongly agreed or agreed.

Participants generally concur (51%) that the ability to remotely control appliances via the energy-saving system could mitigate waning interest over time. However, opinions on this matter differ, as 21% either strongly disagree or slightly disagree with this notion. Additionally, from the gathered data, we identified the primary factors hindering the adoption of sustainable behaviors like energy management systems, namely, the upfront implementation costs, followed by installation complexity and time constraints.

In summary, the survey data indicates a generally favorable perception of energy-saving systems and a readiness to utilize them. Nonetheless, there are individuals who remain uncertain about the long-term utility of these systems. Consequently, it could be advantageous for designers of energy-saving systems to tackle these uncertainties and enhance the user experience to foster adoption and ongoing usage.

The approach demonstrated its ability to sustain user engagement over time, with users actively participating in energy conservation efforts.

Employing swarm intelligence systems can lead to even better savings results due to their ability to harness the power of collective decision-making and adaptability. By mimicking actions taken by other users and replicating cause and effect with either system autonomous interaction with the environment or by giving suggestions directly to users, these systems can effectively navigate complex environments and find optimal solutions to energy management challenges that people alone would not be able to. Individual agents, in this case houses using the saving system, within the swarm could interact locally, exchanging information and responding to changes in the environment or system parameters. This decentralized approach would allow for greater flexibility and robustness in decision-making, as the swarm can quickly adjust its behavior in response to new information or changing conditions. Additionally, swarm intelligence systems often exhibit emergent properties, where complex behaviors and solutions develop from the interactions of simple individual agents, in this case the users themselves with their house-saving systems. This emergent behavior can lead to innovative and efficient strategies for energy conservation, resulting in even greater savings compared to traditional centralized approaches. Overall, by leveraging the principles of swarm intelligence, energy management systems can achieve superior results in terms of both efficiency and effectiveness.

Acknowledgements

This research was partially supported by Fundação para a Ciência e a Tecnologia, IP (FCT) [ISTAR Projects: UIDB/04466/2020 and UIDP/04466/2020 and a PhD scholarship: UI/BD/150751/2020].

The authors thank FCT for providing a PhD scholarship for this chapter's main researcher.

References

Abreu, M.I., de Oliveira, R.A.F. and Lopes, J. (2020). Younger vs. older homeowners in building energy-related renovations: Learning from the Portuguese case. *Energy Reports*, *6* (suppl-1), 159–64. https://doi.org/10.1016/j.egyr.2019.08.036.

Alam, M.R., St-Hilaire, M. and Kunz, T. (2016). Computational methods for residential energy cost optimization in smart grids: A survey. *ACM Computing Surveys*, *49*(1). https://doi.org/10.1145/2897165.

Alshohoumi, F., Sarrab, M., AlHamadani, A. and Al-Abri, D. (2019). Systematic review of existing IoT architectures security and privacy issues and concerns. *International Journal of Advanced Computer Science and Applications*, *10*(7). https://doi.org/10.14569/ijacsa.2019.0100733.

Apanaviciene, R., Urbonas, R. and Fokaides, P.A. (2020). Smart Building Integration into a Smart City: Comparative Study of Real Estate Development. *Sustainability*, *12*(22), 9376. https://doi.org/10.3390/SU12229376.

ASHRAE. (2013). Standard 55-2013: Thermal Environmental Conditions for Human Occupancy. *In*: *ASHRAE*.

Bera, P. (2016). How colors in business dashboards affect users' decision-making. *Communications of the ACM*, *59*(4), 50–57. https://doi.org/10.1145/2818993.

Berardi, U. (2013). Stakeholders' influence on the adoption of energy-saving technologies in Italian homes. *Energy Policy*, *60*, 520–30. https://doi.org/10.1016/j.enpol.2013.04.074.

Böckle, M., Novak, J. and Bick, M. (2020). Exploring gamified persuasive system design for energy saving. *Journal of Enterprise Information Management*. https://doi.org/10.1108/JEIM-02-2019-0032.

Boga, S.R.C., Kansagara, B. and Kannan, R. (2018). Integration of Augmented Reality and Virtual Reality in building information modeling: The next frontier in civil engineering education. *In*: Kurubacak, G, and Altinpulluk, H. (Eds.), (Chapter 12), *Virtual and Augmented Reality: Concepts, Methodologies, Tools, and Applications*. IGI Global. https://doi.org/10.4018/978-1-5225-5469-1.ch049.

Bourikas, L., Gauthier, S., En, N.K.S. and Xiong, P. (2021). Effect of thermal, acoustic, and air quality perception interactions on the comfort and satisfaction of people in office buildings. *Energies*, *14*(2). https://doi.org/10.3390/en14020333.

Caferra, R., Colasante, A. and Morone, A. (2021). The less you burn, the more we earn: The role of social and political trust on energy-saving behaviour in Europe. *Energy Research and Social Science*, *71*, 101812. https://doi.org/10.1016/J.ERSS.2020.101812.

Castaldo, V.L., Pigliautile, I., Rosso, F., Cotana, F., De Giorgio, F. and Pisello, A.L. (2018). How subjective and non-physical parameters affect occupants' environmental comfort perception. *Energy and Buildings*, *178*, 107–129. https://doi.org/10.1016/j.enbuild.2018.08.020.

Chiang, C.T., Chu, C.P. and Chou, C.C. (2015). BIM-enabled Power Consumption Data Management Platform for Rendering and Analysis of Energy Usage Patterns. *Procedia Engineering*, *118*, 554–62. https://doi.org/10.1016/j.proeng.2015.08.480.

Darejeh, A. and Singh, D. (2013). A review on user interface design principles to increase software usability for users with less computer literacy. *Journal of Computer Science*, – (11), 1443–50. https://doi.org/10.3844/jcssp.2013.1443.1450.

Demin, V.F., Krylov, D.A. and Kurachenko, I.A. (2014). Comparison of the damage caused to the environment and public health by NPP and tpp. *Atomic Energy*. Springer. https://doi.org/10.1007/s10512-014-9843-y.

Du, J. and Pan, W. (2021). Evaluating energy saving behavioral interventions through the lens of social practice theory: A case study in Hong Kong. *Energy and Buildings*, *251*, 111353. https://doi.org/10.1016/J.ENBUILD.2021.111353.

García, Ó., Prieto, J., Alonso, R.S. and Corchado, J.M. (2017). A framework to improve energy efficient behaviour at home through activity and context monitoring. *Sensors (Switzerland)*, *17*(8). https://doi.org/10.3390/s17081749.

Geng, Y., Ji, W., Lin, B., Hong, J. and Zhu, Y. (2018). Building energy performance diagnosis using energy bills and weather data. *Energy and Buildings*, *172*, 181–91. https://doi.org/10.1016/j.enbuild.2018.04.047.

Hargreaves, T., Nye, M. and Burgess, J. (2013). Keeping energy visible? Exploring how householders interact with feedback from smart energy monitors in the longer term. *Energy Policy*, *52*, 126–34. https://doi.org/10.1016/J.ENPOL.2012.03.027.

Herrmann, M.R., Costanza, E., Brumby, D.P., Harries, T., das Graças Brightwell, M., Ramchurn, S. and Jennings, N.R. (2021). Exploring domestic energy consumption feedback through interactive annotation. *Energy Efficiency*, *4*, Article No. 90. https://doi.org/10.1007/s12053-021-09999-0.

Hoque, S. and Weil, B. (2016). The relationship between comfort perceptions and academic performance in university classroom buildings. *Journal of Green Building*, *11*(1), 101–117. https://doi.org/10.3992/jgb.11.1.108.1.

IEA. (2022). *World Energy Outlook 2022: Data Product*. www.iea.org/t&c/

Inoue, E., Taniguchi, H. and Yamada, K. (2022). Measuring energy-saving technological change: International trends and differences. *Journal of Environmental Economics and Management*, *115*, 102709. https://doi.org/10.1016/J.JEEM.2022.102709.

Inyim, P., Batouli, M., Reyes, M.P., Carmenate, T., Bobadilla, L. and Mostafavi, A. (2018). A smartphone application for personalized and multi-method interventions toward energy saving in buildings. *Sustainability (Switzerland)*, *10*(6). https://doi.org/10.3390/su10061744.

Isnen, M., Kurniawan, S. and Garcia-Palacios, E. (2020). A-SEM: An adaptive smart energy management testbed for shiftable loads optimisation in the smart home. *Measurement: Journal of the International Measurement Confederation*, *152*. https://doi.org/10.1016/j.measurement.2019.107285.

Ji, Q., Zhang, H.Y. and Zhang, D. (2019). The impact of OPEC on East Asian oil import security: A multidimensional analysis. *Energy Policy*, *126*, 99–107. https://doi.org/10.1016/J.ENPOL.2018.11.019.

Khan, N., Shahid, Z., Alam, M.M., Bakar Sajak, A.A., Mazliham, M.S., Khan, T.A. and Ali Rizvi, S.S. (2022). Energy Management Systems Using Smart Grids: An Exhaustive Parametric Comprehensive Analysis of Existing Trends, Significance, Opportunities, and Challenges. *International Transactions on Electrical Energy Systems*, *2022*. Wiley (online) https://doi.org/10.1155/2022/3358795.

Kummitha, R.K.R. and Crutzen, N. (2017). How do we understand smart cities? An evolutionary perspective. *Cities*, *67*, 43–52 . https://doi.org/10.1016/j.cities.2017.04.010.

LaMarche, J., Cheney, K., Akers, C., Roth, K. and Sachs, O. (2014). Home energy displays: Consumer adoption and response. *In*: *Home Energy Feedback Devices: Adoption and Analyses*, 1–41. Nova Science Publishers, NY, USA.

Li, H., Wang, Z.-H. and Zhang, B. (2023). How social interaction induce energy-saving behaviors in buildings: Interpersonal & passive interactions v.s. public and active interactions. *Energy Economics*, *118*, 106515. https://doi.org/10.1016/J.ENECO.2023.106515.

Liu, Q. (2013). BluePot: An ambient persuasive approach to domestic energy saving. *In*: *Digest of Technical Papers - IEEE International Conference on Consumer Electronics*, *C*, 106–107. https://doi.org/10.1109/ICCE.2013.6486815.

Mataloto, B., Ferreira, J.C. and Cruz, N. (2019). LoBEMS: IoT for Building and Energy Management Systems. *Electronics*, *8*(7), 763. https://doi.org/10.3390/electronics8070763.

Mataloto, B., Ferreira, J.C. and Resende, R.P. (2023). Long-term energy savings through user behaviour modeling in smart homes. *IEEE Access*, *11*, 44544–44558. https://doi.org/10.1109/ACCESS.2023.3272888.

McIlvennie, C., Sanguinetti, A. and Pritoni, M. (2020). Of impacts, agents, and functions: An interdisciplinary meta-review of smart home energy management systems research. *In*: *Energy Research and Social Science*, *68* (*October*, 101555. https://doi.org/10.1016/j.erss.2020.101555.

Mi, L., Qiao, L., Gan, X., Xu, T., Lv, T., Qiao, Y. and Ding, C. (2020). Assessing the effect of non-financial information intervention on promoting group-level energy savings. *Science of the Total Environment*, *720* (June), 137533. https://doi.org/10.1016/j.scitotenv.2020.137533.

Nižetić, S., Djilali, N., Papadopoulos, A. and Rodrigues, J.J.P.C. (2019). Smart technologies for promotion of energy efficiency, utilization of sustainable resources and waste management. *Journal of Cleaner Production*, *231*, 565–91. https://doi.org/10.1016/J.JCLEPRO.2019.04.397.

Onuma, H., Matsumoto, S. and Arimura, T.H. (2020). How much household electricity consumption is actually saved by replacement with Light-Emitting Diodes (LEDs)? *Economic Analysis and Policy*, *68*, 224–38. https://doi.org/10.1016/J.EAP.2020.09.015.

Perea-Moreno, A.J. (2021). Renewable Energy and Energy Saving: Worldwide Research Trends. *Sustainability*, *13*(23), 13261. https://doi.org/10.3390/SU132313261.

Qu, J. (2022). Research on energy saving of computer rooms in Chinese colleges and universities based on IoT and edge computing technology. *Heliyon*, *8*(10), e10970. https://doi.org/10.1016/J.HELIYON.2022.E10970.

Ramya, C.M., Shanmugaraj, M. and Prabakaran, R. (2011). Study on ZigBee technology. *In*: *ICECT 2011–2011 3rd International Conference on Electronics Computer Technology*, *6*, 297–301. https://doi.org/10.1109/ICECTECH.2011.5942102.

Reddy, A.R. and Battu, J.R. (2019). An Analytical Study on Air Pollution: Perspective of the Environmental. *Think India*, *22*(4). https://doi.org/10.26643/think-india.v22i2.8757.

Roy, R. and Caird, S. (2007). Designing low and zero carbon products and systems: Improvements based on consumers' experience of adoption and use. *In*; *Proceedings of ICED 2007, the 16th International Conference on Engineering Design*, *DS 42*.

Ruokamo, E. (2016). Household preferences of hybrid home heating systems: A choice experiment application. *Energy Policy*, *95*, 224–37. https://doi.org/10.1016/j.enpol.2016.04.017.

Salamone, F., Belussi, L., Currò, C., Danza, L., Ghellere, M., Guazzi, G., Lenzi, B., Megale, V. and Meroni, I. (2018). Application of IoT and Machine Learning techniques for the assessment of thermal comfort perception. *Energy Procedia*. https://doi.org/10.1016/j.egypro.2018.08.130.

Shahinmoghadam, M., Natephra, W. and Motamedi, A. (2021). BIM- and IoT-based virtual reality tool for real-time thermal comfort assessment in building enclosures. *Building and Environment*, *199*, 107905. https://doi.org/10.1016/j.buildenv.2021.107905.

Simanaviciene, Z., Volochovic, A. and Simanavicius, A. (2014). Analysis of energy efficiency improvement and climate change mitigation policy in Lithuania. *WIT Transactions on Ecology and the Environment*, *186*, 739–49. https://doi.org/10.2495/ESUS140661.

Soomro, A.M., Bharathy, G., Biloria, N. and Prasad, M. (2021). A review on motivational nudges for enhancing building energy conservation behavior. *Journal of Smart Environments and Green Computing, 2021*. https://doi.org/10.20517/jsegc.2020.03.

Visconti, P., Costantini, P., De Fazio, R., Lay-Ekuakille, A. and Patrono, L. (2019). A sensors-based monitoring system of electrical consumptions and home parameters remotely managed by mobile app for elderly habits' control. *In*: *Proceedings - 2019 8th International Workshop on Advances in Sensors and Interfaces, IWASI 2019*, 264–69. https://doi.org/10.1109/IWASI.2019.8791399.

Weerasinghe, A.S., Rotimi, J.O.B. and Rasheed, E.O. (2023). Modelling of underlying social psychological effects on occupant energy-related behaviours. *Building and Environment*, *231*, 110055. https://doi.org/10.1016/J.BUILDENV.2023.110055.

Wemyss, D., Cellina, F., Lobsiger-Kägi, E., de Luca, V. and Castri, R. (2019). Does it last? Long-term impacts of an app-based behavior change intervention on household electricity savings in Switzerland. *Energy Research and Social Science*, *47* (August 2018), 16–27. https://doi.org/10.1016/j.erss.2018.08.018.

Wut, E., Ng, P., Leung, K.S.W. and Lee, D. (2021). Do gamified elements affect young people's use behaviour on consumption-related mobile applications? *Young Consumers*. https://doi.org/10.1108/YC-10-2020-1218.

Xu, D., Sheng, M., Zhao, X., Qian, H., Liang, W., Zheng, X. and Zhao, D. (2022). Electrochromism-induced adaptive fresh air pre-handling system for building energy saving. *Energy and Built Environment*, *5*(2), 300–308. https://doi.org/10.1016/J.ENBENV.2022.10.005.

Yang, J., Wu, J., Xian, T., Zhang, H. and Li, X. (2022). Research on energy-saving optimization of commercial central air-conditioning based on data mining algorithm. *Energy and Buildings*, *272*, 112326. https://doi.org/10.1016/J.ENBUILD.2022.112326.

Yera, A., Muguerza, J., Arbelaitz, O., Perona, I., Williams, R., Peek, N., Keers, R., Ashcroft, D., Jay, C. and Vigo, M. (2018). Inferring visual behaviour from user interaction data on a medical dashboard. *In*: *ACM International Conference Proceeding Series*. https://doi.org/10.1145/3194658.3194676.

7 Swarm Intelligence for Building Energy Management

An Overview of the Algorithms Used for Different Types of Energy Needs

Leonor Domingos[1*]

1. Introduction

Decarbonizing the energy system and moving toward net zero energy have been major topics of discussion and policy in the global energy arena. The necessity of making a swift transition to a net-zero future is made abundantly evident by the ongoing increase in carbon emissions and the rising frequency of extreme weather events in recent years (BP Energy Economics, 2023). The building and construction sector was responsible for about 39% of all carbon emissions worldwide in 2018, and accounts for about 36% of world-wide energy consumption. If demolition is also accounted for, then the number rises to almost 50% (Santamouris and Feng, 2018). Prospective scenarios on building energy consumption suggest strategies to reduce the sector's energy usage by 2050, or later, by utilizing clean power extensively and enhancing energy-saving measures within structures. Achieving this goal will cost a significant increase in investments and the adoption of strict green policies. Globally, increasing the building sector's energy efficiency is an essential objective. In addition to reducing energy consumption, increased building energy efficiency raises living standards and thermal comfort. Furthermore, energy efficiency increases industrial development, creates new jobs, and has a high added value for economies. The creation of new goods and services that boost the effectiveness of a building's exterior, the energy efficiency of appliances, HVAC, lighting, and other systems, as well as the utility of control systems, are all related to energy efficiency (Santamouris and Vasilakopoulou, 2021). However, the construction industry's energy efficiency is not keeping up with the agreed-upon criteria for decarbonization, as outlined by the International Energy Agency (IEA). Energy intensity has increased by 1.2% in 2018, a significant decrease from the 1.7% and 2.5% levels in 2017 and 2000, respectively, and well below the 3% efficiency improvement pace that was agreed upon to decarbonize the industry (IEA, 2019). The building industry tends to be resistant to technological

[1] Department of Applied Digital Technologies, ISCTE – University Institute of Lisbon.
* Corresponding author: Leonor_domingos@iscte-iul.pt

advancements, although a few technologies offer major advances, but progress is quite slow.

2. Swarm Intelligence: Concepts and Applications

Swarm intelligence (SI) is a specific area within the larger field of artificial intelligence (AI) and is predicated on how components in decentralized and self-organized systems behave together. These systems took inspiration from swarms of insects/animals which collaborate (Kumar, Chatterjee, Payal, and Rathore, 2022). Natural insect swarms, within a determined environment, comprising of simple individuals, can act locally with one another and their surroundings, to solve complex problems (such as ant/bee colonies). The agents of these colonies seem to follow specific rules with no centralized control. The somewhat random interactions among them result in 'intelligent' global behavior, which the individual agents are unaware of. This concept is then translated into a technological environment. SI was first introduced in the 1980s and, since then the scientific community has given it more attention and adapted it into various domains, such as engineering, construction, computer science, and more. Concurrently, an extensive array of SI methods have been developed and experimented to address intricate issues such as optimization challenges (Iglesias, Gálvez, and Suárez, 2020).

3. Energy Distribution in Cities and Smart Grids

There are three energy distribution conundrums in developed cities energy grid concepts. The first is the difficulty of balancing production and demand at the grid level, due to the amount of energy required by cities and buildings to function. The second one relates to the complex processes of transforming the right amounts of thermal energy as it's necessary for comfort. Lastly, the third one relates to the inability of buildings to communicate with one another, as energy-using entities, for optimized energy usage and sharing (Esfandi et al., 2024). Clustering behavior models, such as those supported by SI, could be used to optimize buidling's energy usage, ensuring everything is running smoothly, and then having those agents share information with one another, for maximum optimization. These models can be implemented in smart grids, which together with decentralized energy systems, particularly in the context of urban sustainability, have immense transformative potential (Alotaibi, Abido, Khalid, and Savkin, 2020). These systems leverage advanced technology such as data analytics, machine learning, and the Internet of Things (IoT) devices to optimize energy distribution, enhance resilience, and integrate renewable energy resources. These systems also promote energy autonomy, efficiency, and community engagement. However, these systems also pose challenges regarding complexity, cybersecurity risks, high implementation costs, and require comprehensive solutions, inclusive technological transitions, and community-centered approaches to ensure the full potential of smart grids in advancing sustainable urban energy systems.

4. Methods

To conduct this literature review, an integrated bibliometric analysis and systematic literature review is used. As stated in literature, critiques are important to gather and evaluate the body of information in a determined field, posing issues that cannot be considered by individual studies alone, pointing out flaws in primary research, and advancing the creation or assessment of ideas (Page et al., 2021). For the review, the Preferred Reporting Items for Systematic Reviews and Meta-Analysis (PRISMA) framework is used, to ensure transparency, completeness, and consistency in reporting findings. Primary scientific sources from the Scopus and Web of Science database are considered, the two largest abstract and citation databases, covering a wide range of disciplines, with a comprehensive coverage of scientific literature. a search on both databases for Article title, abstract, and keywords, considering the key terms 'Energy', 'Efficiency', 'Swarm', 'Intelligence', and 'Buildings' is conducted, for all years up to 2024, where 67 documents are found. Then, these documents are screened using the PRISMA framework, considering inclusion and exclusion criteria, which can be seen in Table 1.

Table 1 Inclusion and exclusion criteria for the PRISMA framework.

Inclusion Criteria	*Exclusion Criteria*
Within Scope: Papers that address topics related to the energy management of buildings using swarm intelligence	Out of Scope: Papers that focus on swarm intelligence in other fields
Relevance: Papers that consider swarm intelligence for heating and/or cooling and/or the electrical systems of buildings	Irrelevance: Papers focusing on using swarm intelligence for other aspects of building management
Time frame: Papers published until 2024	Non-English language papers
Publication type: Peer-reviewed journal articles and conference proceedings	Publication type: Non-peer review articles and reports

Source: Author's own elaboration.

Of the 67 initial articles, 15 duplicates were removed, which resulted in 52 articles to screen. The tiles and abstracts of these articles were screened taking into consideration the criteria presented in Table 1. This resulted in a final selection of 22 studies that were analyzed according to the type of building energy requirements that the research focused on. These types and results are presented in the following section.

For the bibliometric analysis, the total 67 papers are considered and the software VOSviewer is used. This is a free software which allows for the visualization of keyword relationships, by using co-occurrence data. This provides insight into the relationships between key terms, allowing researchers to comprehend trends and connections between research fields (Eck and Waltman, 2010).

5. Results and Discussion

5.1 Bibiometric Analysis

Using VOSviewer software, the co-occurrence map (Figure 1) creates a visual story of the research field by displaying keywords, clusters, and their interactions. The size of each node in this map, which represents a distinct keyword, indicates how frequently or significantly that keyword appears in the literature (Tsilika, 2023). The relationships between nodes indicate themes that are frequently addressed together and show frequent co-occurrences. The degree of correlation between terms is indicated by the thickness and proximity of these links. Nodes that are close to one another on the map have a greater tendency to co-occur, which indicates a closer bond between them. Furthermore, the terms within three clusters show stronger correlations with one another than with keywords from other clusters. Thematic coherence of terms in the same cluster indicates that they are frequently discussed alongside or in connection with one another.

Three clusters in the co-occurrence map present the same number of items. Cluster 1 is presented in red, cluster 2 in green, and cluster 3 in dark blue. These are followed by cluster 4 in yellow, cluster 5 in purple, cluster 6 in light blue, and cluster 7, with just one item, in orange. The first cluster focuses on themes regarding cooling and heating systems and heat storage, thus focusing on the comfort systems of buildings, as well as networks, more specifically artificial neural networks, data mining digital storage, and sensitivity analysis. This connection relates to the use of information for the adequate management of buildings and their inhabitants' comfort. The second cluster is focused solely on energy resources and management of buildings, while still considering the possibility of integration with more advanced digital systems, through learning systems and machine learning. Cluster 3 is focused on sustainable development, architectural design, the construction industry, environmental impact, renewable technology, and AI. While this cluster is mainly focused on the impact of buildings in the environment, it keeps a connection with AI, as it can contribute to architectural design using generative algorithms for the planning and development of buildings, and this includes SI (Bölek, Tutal, and Özbaşaran, 2023). Cluster 4 is focused on sensors, networks, algorithms, swarm particle optimization (SPO), SI, and quality control for energy efficiency. This cluster shows the interactions between sensors and networks for energy efficiency, considering the usage of SI algorithms. Cluster 5 shifts the focus to the built environment as a whole, considering the scope of smart cities, through focus on intelligent (smart) buildings, smart grids, smart homes, and power grids. This cluster is more related to the urban city environment as opposed to just the building as a specific entity. Cluster 6 is solely focused on smart buildings and smart cities, as more general concepts, without connection to energy specifically or a determined set of technologies. Lastly, cluster 7 consists of only one item, indicating its distance and disconnection from most other terms, which relates to robotics. This means that in terms of research there's a need and a greater potential

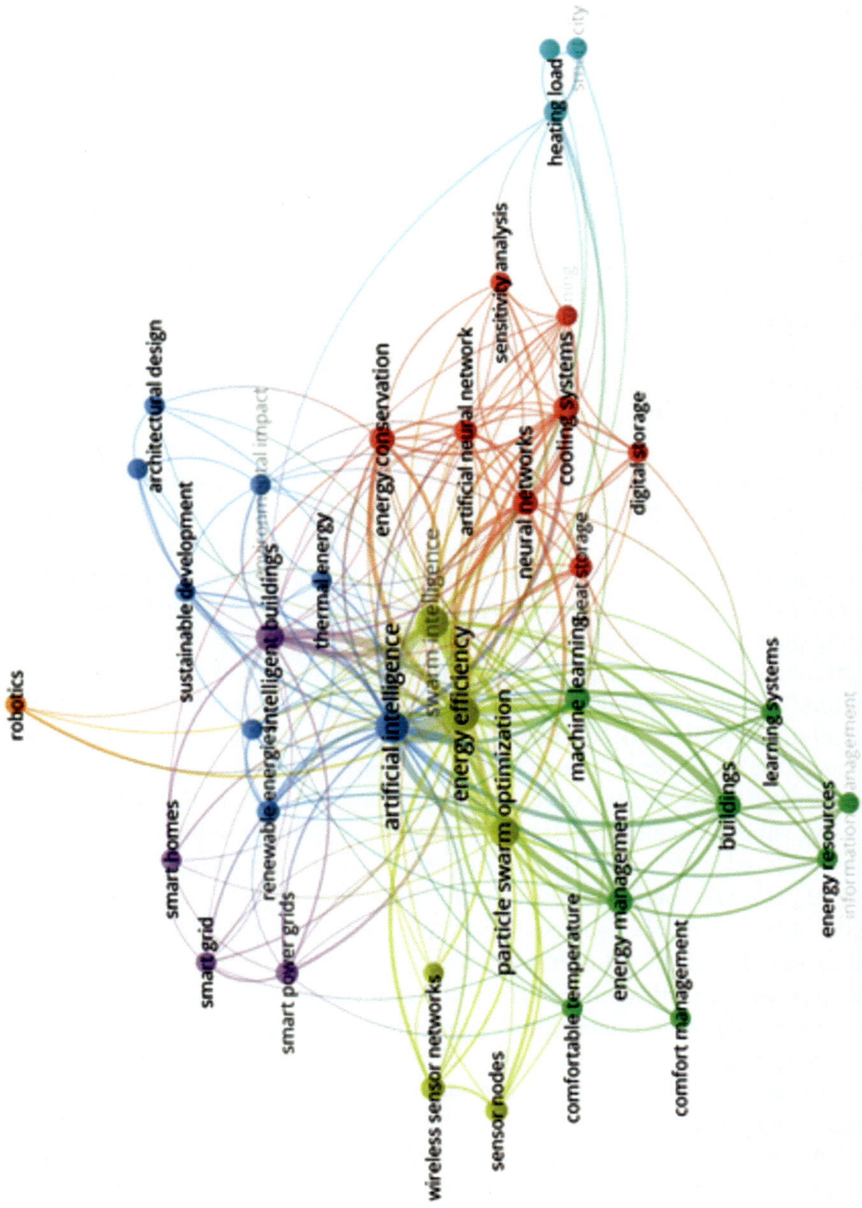

Fig. 1 Bilbiometric Analysis considering the research key terms (Author's own elaboration, using the VOSviewer software).

to further explore this theme in correlation with the rest, for the application of SI in the energy management of buildings.

Figure 1 offers information that both validates current understanding and suggests new lines of inquiry. A significant density of connected nodes, signifying thoroughly researched subjects, can be seen in certain regions of the map. Some places, on the other hand, have relatively few links, suggesting that they have not yet undergone thorough investigation. However, even within the highly connected clusters that represent well-established fields of research, there could be unresolved issues or gaps in knowledge that the mainstream academic community has overlooked.

Regarding the number of documents produced per year, it's possible to see in Figure 2 that the documents retrieved from the research first appeared in 2008, and the number has been rising since, with the highest total of documents being published in 2023, with eight documents. This highlights that there has been development in the field but there's a great potential for further exploration.

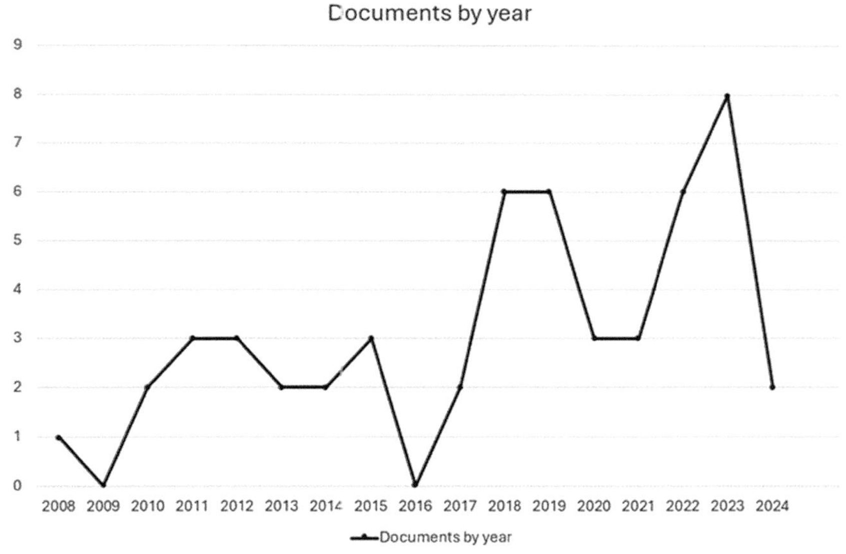

Fig. 2 Documents produced by year, from 2008 to 2024 (Author's own elaboration).

Figure 3 presents the documents by subject area, considering the total number of documents gathered, before going through the criteria present in Table 1. It's possible to see that there's still much research to be done in the fields of "engineering" (28.3%), and "energy" (15%), although engineering is the subject area where the field has been more explored, followed directly by "computer science" (21.2%). Both "environmental science" and "Social sciences" have the same number of documents (6%), followed by "Chemical Engineering", "Material Science", "Earth and Planetary Sciences", "Physics and Astronomy", and other fields, each below 3%. Since the research included the term "Buildings", it's comprehensive that the latter subjects show less number of documents than the first ones, related to engineering and energy.

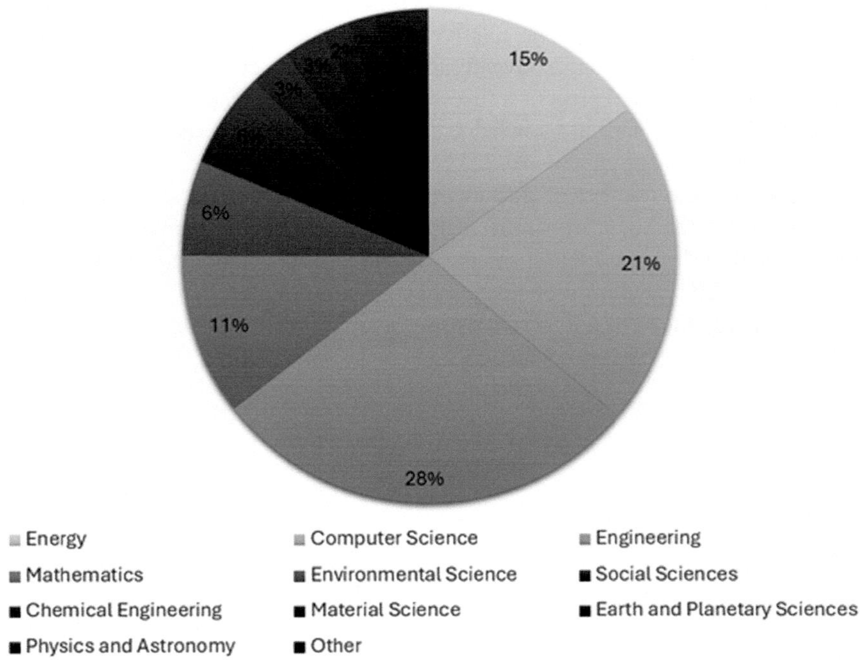

Fig. 3 Documents by subject area (Author's own elaboration).

Lastly, Figure 4 shows the documents by country or territory, allowing one to comprehend where this type of research is being done geographically. It's possible to see that most documents have been produced by Chinese universities, followed by the United States and Vietnam, but with a great difference (20 documents to 8). India is next, followed by Germany with 4 documents. Germany is the only European country with more than 2 published documents. This shows that by a very large margin, most of the research regarding swarm intelligence and energy management for buildings is being developed in Asia.

5.2 Literature Review Analysis

Of the 67 initial articles, 15 duplicates were removed, which resulted in 52 articles to screen. The titles and abstracts of these articles were screened taking into consideration the criteria presented in Table 1, which resulted in a final selection of 22 studies. The results of these studies were analyzed. In the first stage, the research was focused on the type of building energy usage. First, the articles

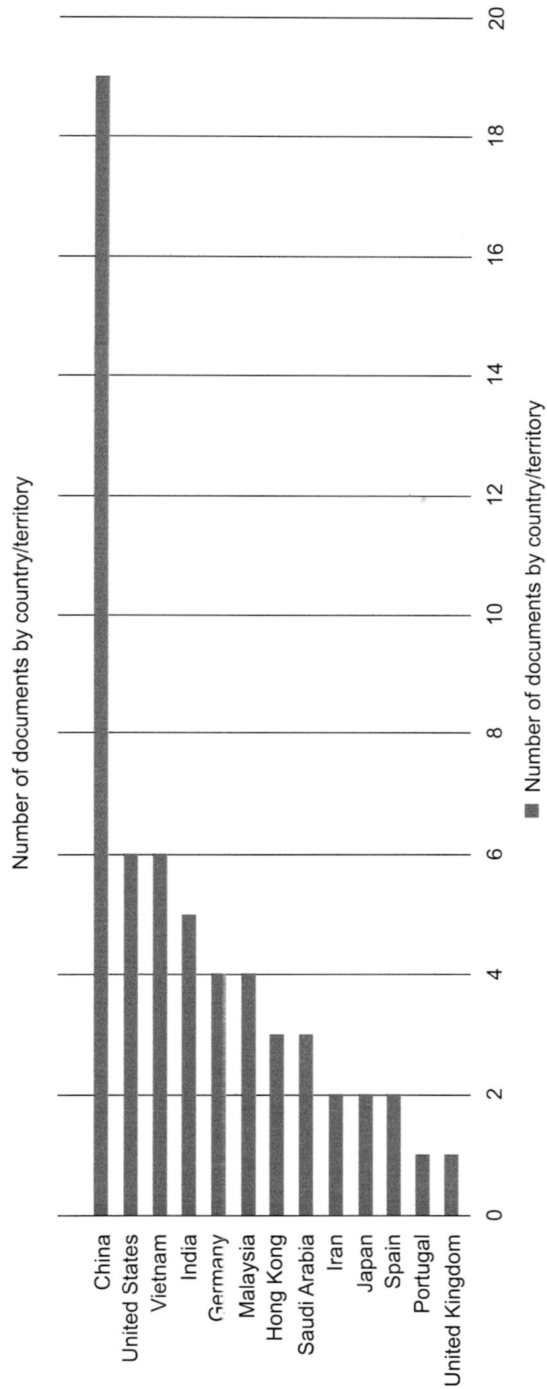

Fig. 4 Number of documents by country or territory (Author's own elaboration).

regarding general energy savings for buildings were analyzed, followed by the articles that considered CO^2 Concentrations, Heating Load, Cooling Load, HAVAC Systems, and Thermal Energy Storage. Lastly, there is one research on Building Information Modeling (this isn't a type of building energy requirement, but refers to the utilization of specific software for energy calculations) which was also analyzed. Secondly, the types of algorithms used for the type of building energy requirements were also indicated, to allow an overview of the various types of SI algorithms being used for building energy management. The results of this analysis are presented below.

5.2.1 Energy savings for smart buildings

The International Energy Agency (IEA) indicates that, worldwide, energy consumption has increased by 92% since 2015, as it's necessary to keep interior environments comfortable for users in terms of air quality and thermal comfort (IEA, 2022). However, even with large amounts of energy being spent, research indicates that only about 80% of buildings' users, in 26% of buildings, are content with the indoor comfort conditions they're being offered (Ortiz, Carrere, Salom, and Novoa, 2023). This is because indoor comfort control systems are inefficient, and unable to adequately predict the ideal conditions for the inhabitants comfort (Kajjoba, Kasedde, Olupot and Lwanyaga, 2022). AI algorithms could be used to adequately predicting the conditions required for indoor comfort; however, these must consider air quality, ideal thermal conditions for human comfort, and ideal energy savings (Ngarambe, Yun and Santamouris, 2020). The documents analyzed in this literature review propose various solutions optimized by SI and AI, with various goals, depending on buildings' energy needs, which ultimately contribute to the energy savings of smart buildings.

Zhang et al. (2024) proposed a new forecasting system (behavior model) for Air Conditioning (AC) energy consumption, for five different buildings, considering the number of occupants, to improve the already-existing Particle Swarm Optimization (PSO) and Grey Wolf Optimization (GWO) algorithms. The GWO algorithm is a nature-inspired optimization technique that mimics the social hierarchy and hunting behavior of gray wolves in the wild, by mimicking the hierarchical leadership and hunting strategies of gray wolves. It has been successfully applied to a wide range of optimization problems, including engineering design, machine learning parameter tuning, feature selection, and scheduling problems (Dada et al., 2022). This new forecasting system measures the combined usage of all AC machines in a building, and unifies heating, cooling, and transition seasons. The probability model can also consider aspects such as the different comfort feelings among different occupants, by analyzing the input from different AC devices, and managing the energy consumption. They named this novel parameter Group occupant behavior probability (GOBP). Using GOBP and a new type of neural network called Long Short-Term Memory (LSTM), which can be optimized using either PSO or GWO, the authors were able to improve the

Mean Absolute Percentage Error (MAPE) from 20.08% to 78.37% across the five buildings used for case study. However, although it improves prediction accuracy, using this model requires a much longer time to compute than previously used general systems. The study also concluded that GOBP might not be suitable for buildings with centrally controlled ACs and uncertain administrator behaviors. This suggests that the effectiveness of GOBP depends on the specific context and characteristics of the building and its systems (Zhang, Ma, Han and Zhao, 2024).

5.2.2 CO2 concentrations

L. Li et al. (2023) proposed an adaptive multi-objective optimization algorithm, which involves the resolution of problems through mathematical optimization, considering multiple-criteria decision-making (Chang, 2015), and incorporates a backpropagation neural network (BPNN), where input data is fed through a network, in a forward direction, to generate a determined output (Anand, Mishra and Khan, 2022). This model allows to predict and optimize the indoor quality of buildings, as well as temperature comfort and, as such, guarantee energy efficiency. This is achieved by first creating a database of the indoor air flow and external temperature, from which the BPNN model predicts CO2 concentration and the indoor thermal comfort, to ensure smart interior environmental management. In case these values are not within the threshold of what is considered comfortable for the buildings' inhabitants, then an adaptive multi-objective GWO is used to identify the best combination of parameters to ensure indoor air quality, thermal comfort, and energy savings (Li, He, Zhang, Fung and Lau, 2023). They concluded that the model could predict CO2 concentrations with an accuracy of more than 90%, and optimize concentrations by up to 45%, with energy savings of 35%. Also, the use of SI allowed reduction of the runtime of the algorithms by 98.4%, when compared to computational fluid dynamics (CFD) simulations.

M. Hossain et al. (2023) used the principles of green building, which is the practice of designing buildings and utilizing resource-conscious and efficient techniques across a building's whole life cycle—from choosing the location to planning, developing, running, maintaining, and renovating (Bungau, Bungau, Prada and Prada, 2022), to develop two hybrid neural networks. The purpose of these algorithms is to use constraint optimization techniques to minimize the number of impractical solutions, hence improving the classifier output. The output of a conventional backpropagation neural network serves as the input for the hybrid neural network algorithm (Parikh, DaPonte and Vitale, 1999). These networks rely on a variety of algorithms, specially Backtracking Search Algorithms (BSA), which is a population-based evolutionary algorithm designed for solving optimization problems, and works by evolving a population of candidate solutions through processes of mutation, crossover, and selection (Chen, Zou, Lu and Li, 2019); Cuckoo Optimization Algorithms (COA), which is inspired by the brood parasitism of some cuckoo species which lay their eggs in the nests of other birds, to optimize solutions (it involves generating new solutions, cuckoo eggs, selecting the best solutions, and replacing the worst solutions, mimicking the survival of the fittest) (Guerrero-Luis, Valdez and Castillo, 2021); Artificial Neural Networks (ANNs), which are computational models inspired by the way biological neural networks in the human brain work, are widely used for pattern recognition, classification, regression, and other machine learning tasks (López, López and Crossa, 2022); BSA-MLP, which is the combination between BSA and Multilayer Perception, another type of ANN, improves tasks such as classification or regression, this leverages the optimization power of BSA to enhance the learning capability of the MLP (Moayedi et al., 2023); and lastly, COA-MLP, which uses COA to optimize the parameters of the MLP, aiming to achieve better accuracy and generalization in the neural network's predictions (Abdolrasol et al., 2021). The authors concluded that their approach allowed to effectively estimate global carbon dioxide emissions of various types of building walls, namely reinforced concrete, red brick, aerated concrete, and briquette. Of all the utilized algorithms, the COA showed the best outcomes for calculating the total heat transfer of the same walls, to account for energy efficiency. This approach allowed for the comparison of both the carbon emissions and energy efficiency of materials for residential buildings' walls, and these models may contribute to optimize building design (Moayedi, Yildizhan, Al-Bahrani and Van, 2023).

5.2.3 Heating load

The heating load of buildings is an essential part of energy efficiency, as well as a building's energy consumption. This makes it a critical question when optimizing smart buildings and smart cities planning, ensuring a more sustainable future for the built environment and those who live in it.

Zhou, Zheng, and Zhang (2019) proposed a new, combined, renewable energy system for buildings, which consists of two main parts, a ventilated Trombe wall (which is a type of solar heating system which harnesses solar energy to heat an interior space), with phase change materials (PCMs-VTW) and a photovoltaic setup, both for heat storage. Phase Change Materials (PCMs) are substances that absorb or release a significant amount of latent heat when they change their physical state, such as from solid to liquid or vice versa. This property allows them to store and release thermal energy efficiently, making them useful in various applications for temperature regulation and energy storage (Hossain et al., 2023). The heat transfer process of these systems was maximized using an optimization methodology, which implemented machine learning-based algorithms, and the development of an ANN-based data-driven learning algorithm. To maximize the overall energy generation, both teaching-learning-based optimization (TLBO) and PSO are used. Between both algorithms, the TLBO resulted in a solution that provided more cooling energy but with a thinner PCM storage layer compared to the solution from the PSO algorithm, which is a recommended solution for practical engineering applications, and thus a better practical solutions for system design (Zhou, Zheng and Zhang, 2019).

Nguyen, Zhou and Dou (2019) developed and proposed four hybrid models based on AI to estimate the Heating Load of Energy Efficient Building systems. The authors simulated 12 different types of buildings in the Ecotect computer software and used 768 experimental datasets. They developed the following types of hybrid computational models which combine ANN and other optimization algorithms. GA-ANN, which combines genetic algorithms (GA), search heuristics inspired by the process of natural selection, which use techniques such as selection, crossover, and mutation to evolve solutions to optimize problems (Albadr, Tiun, Ad-Dhief and Ayob, 2020), and ANN. In this hybrid model, GA is used to optimize the weights and biases of the ANN. The GA iterates through generations of potential solutions, selecting and breeding the fittest individuals to find an optimal set of ANN parameters. This process helps to avoid local minima and improve the overall performance of the ANN. PSO-ANN, which combines PSO and ANN, where PSO is employed to optimize the weights and biases of the ANN. Each particle represents a potential solution (a set of ANN parameters), and particles adjust their positions based on their own experience and the experience of their neighbors. This collective search strategy helps in efficiently finding optimal or near-optimal ANN configurations. ICA-ANN, where an Imperialist Competitive Algorithm (ICA) is used in combination with ANN. ICA is an optimization technique based on the socio-political process of imperialistic competition. It models the competition among empires to take over colonies, aiming to find the optimal solution (Emami and Parsa, 2020). Combining both ICA and ANN optimizes the parameters of ANN, the algorithm starts with a population of possible solutions (countries), and these countries form empires. Over iterations, weaker empires are taken over by stronger ones, converging towards an optimal set of ANN parameters. Lastly, ABC-ANN, where Artificial Bee Colony (ABC), which is an optimization algorithm inspired by the foraging behavior of honeybees, and involves three types of bees: employed bees, onlookers, and scouts, which collaborate to find the best solution (Xiao, Li and Tan, 2023), is joined with an ANN, optimizing it. Employee bees explore the solution space based on their current knowledge, onlookers choose solutions based on the quality of food sources (solution quality), and scouts randomly search new areas. This approach helps in effectively finding optimal or near-optimal ANN configurations. Of all the algorithms used in the study, the GA ensured the highest performance in optimizing the ANN model to predict heating loads (GA-ANN). The rest of the algorithms showed less satisfactory performances. Either way, the study showed that the use of these algorithms can adequately predict (and later manage) the energy efficiency in buildings in smart cities (Nguyen, Zhou and Dou.

Moayedi, Nguyen and Foong (2021) also used a Grasshopper optimization algorithm (GOA), which is a metaheuristic optimization technique inspired by the swarming behavior of grasshoppers in nature (it starts with an initial population of grasshoppers, each representing a potential solution to the optimization problem) (Wu, Wu and Wang, 2023), and GWO to improve the neural assessment of the heating load of residential buildings. The researchers coupled these algorithms with MLP to find the best settings for the neural network. It was concluded that using population sizes of 300 for GOA-MLP and 400 for GWO-MLP gave better results than other configurations, while the GWO method took more time to compute than the GOA method, and while both methods improved the MLP's predictions, GWO did a better job overall than GOA, although it took longer to complete (Moayedi, Nguyen and Foog, 2021).

5.2.4 Cooling load

Cooling load refers to the amount of heat that must be removed from a space to maintain a desired temperature and humidity level. It is a crucial factor in the design and operation of air conditioning and refrigeration systems.

Nedjah, Mourelle, and Lizarazu (2022) explored the use of SI for the operation of cooling towers; in conjunction with chillers, these towers are often used in large-commercial buildings and power plants in the refrigeration systems. The authors explored the use of both GA and PSO algorithms, and created two system models, one based on a single search variable, and one on two search variables. It was concluded that the model with two search variables and the PSO algorithm ensured the best results. This scenario achieved a cooling efficiency factor of 1.78, resulting in energy savings of 9.48%, with a 5.32% reduction in tower effectiveness (Nedjah, Mourelle and Lizarazu, 2022).

Nguyen et al. (2019) employed PSO algorithms to assess the environmental impact of the cooling load in energy-efficient building systems and examined four machine learning models and their optimization using PSO. The objective of the research was to introduce the most reliable method to estimate a building's cooling load. Although the machine learning models have decent estimation results, the hybrid models using the PSO algorithm are more reliable and accurate (Nguyen, Moayedi, Jusoh and Sharifi, 2020).

5.2.5 HVAC systems

HVAC stands for Heating, Ventilation, and Air Conditioning. An HVAC system is a technology used to control and regulate indoor environmental conditions, including temperature, humidity, air quality, and ventilation, within a building (Asim et al., 2022).

Moayedi and Mosavi (2021) studied the simultaneous calculations of heating and cooling loads of residential buildings through the HVAC system. They used the following algorithms: GOA; wind-driven optimization (WDO), which is a nature-inspired optimization algorithm based on the behavior of wind particles; as a population-based metaheuristic optimization algorithm, it does not require gradient information to search for optimal solutions (Tomar, Bansai and Singh, 2024); and biogeography-based-optimization (BBO), like WDO, it is also a nature-inspired optimization algorithm but based on the principles of biogeography, the study of the distribution of organisms in different environments over time (Lim, Wibowo, Desa and Haron, 2016). Their study mixes these algorithms with ANNs and MLPs, concluding that the accuracy of the BBO-based model was slightly higher. The algorithm predictions were very close to the real values of the energy loads and can be later adapted to use the data of other types of buildings (Moayedi and Mosavi, 2021).

Garces-Jiménez et al. (2021) compared the performance of various swarm-based algorithms in optimizing the HVAC system, using the Teatro Real de Madrid as a case study. The authors used multi-objective genetic algorithms (MOGA), Non-Dominated Sorting Genetic Algorithm II (NSGA-II), NSGA-III (an extension of the previous) which are all a variety of GA, Multi-Objective Particle Swarm Optimization (MOPSO), which is a variant of PSO, and Random Search, a simple optimization technique that involves randomly sampling potential solutions from the search space and evaluating their performance using an objective function (Zou, Liu, Wang, Li and Wu, 2021). The research concluded that the GAs perform faster than MOPSO and Random Search, although MOPSO has better convergence metrics. They achieved energy savings from 60 to 80% (Garces-Jimezes, Gomez-Pulido, Gallego-Salvador and Garcia-Tejedor, 2021).

5.2.6 Thermal Energy Storage (TES)

Thermal energy storage (TES) is related to the capacity to temporarily store and reallocate thermal energy, and has been applied to various industries, such as power plants or wind farms (Elkhatat and Al-Muhtaseb, 2023). It opens various possibilities for ensuring renewable energy usage and insures energy savings. Combining TES with AI and algorithms is adamant to gain economic and environmental benefits.

He, Guo, and Zhang (2022) produced an overview of AI in TES, such as predicting performance and characteristics, optimizing design, and controlling operation. These techniques are applied across different scenarios, including domestic water heaters, HVAC systems in buildings, large-scale energy systems, and industrial setups. AI techniques are highlighted for their ability to handle uncertain and imprecise factors effectively and at a low cost. Specifically mentioned methods include ANN, GA, PSO, and Fuzzy Logic (FL), which is a mathematical approach used to handle reasoning with imprecise or uncertain information, and allows for degrees of truth between one and 0, as opposed to binary logic (Zadeh, 2015), and combinations thereof. These AI methods are noted for their capability to work even when detailed physical knowledge is lacking, helping in accurate estimation of TES performance, system optimization, and efficient energy distribution to meet various objectives like demand satisfaction, cost savings, and environmental impact reduction. However, limitations of these technologies are highlighted, AI cannot directly reveal unknown physical knowledge and mechanisms underlying TES systems. Additionally, the lack of sufficient data can hinder the establishment of accurate prediction models for TES (He, Guo and Zhang, 2022).

On a smaller scale, Li et al. (2024) analyzed a Passivhaus building, an example of a green, environmental protective, low-carbon, and energy-saving ultra-low energy building (Hafez, Sa'di, Gamal and Taufiq-Yap, 2023) with a prefabricated frame structure, located in the eastern part of Jinan City, with a typical temperate monsoon climate. Near-zero-energy buildings are essential for the development of building energy efficiency in cities. Their study proposed a new approach for thermal energy storage, while leveraging passive house construction. They were able to reduce the energy consumption of the model by 19.24% and 19.70%, depending on the algorithm that was used. The authors used GenOpt, an open-source optimization program used primarily for optimizing energy systems and building performance. It is designed to find the optimal values of input parameters to minimize or maximize an objective function, typically related to energy consumption, cost, or environmental impact (Serbouti, Rattal, Abdellah, Harmouchi and Mouhsen, 2018). To achieve optimization of their system, they used a PSO algorithm with Inertia Weight (PSOIW), which is variant of the classic PSO. By introducing an inertia weight factor into the velocity update equation of the PSO algorithm, it controls the balance between exploration and exploitation during the optimization process. It essentially determines how much a particle's current velocity influences its future velocity. A higher inertia weight encourages exploration, allowing particles to move more freely in the search space, while a lower inertia weight promotes exploitation, causing particles to converge towards promising regions of the search space (Liu, Mei and Li, 2015). Regardless of the study's results, the approach is only theoretical, and requires validation experiments for cooling and heating seasons to enhance the model's precision (Li et al., 2024).

5.2.7 Electrical systems

Electricity is the main source of worldwide use of energy, for various uses, and is a necessity for smart buildings and cities. The IEA expects a global increase of 28% in electricity use until 2040, mainly in residential and commercial buildings (IEA, 2022).

Malik and Kim (2018) used PSO algorithms to understand energy usage to optimize energy consumption in smart buildings in Seoul, South Korea. The datasets were gathered during the timespan of one year. The authors used two variations of PSO algorithms, one which incorporates a mechanism for regenerating updating particles (solutions) and another which incorporates a mechanism for boosting the velocity of particles in the algorithm. The PSO algorithm with the velocity boost showed higher accuracy, while the algorithm with the regeneration also achieved better accuracy than the regular PSO algorithm (Malik and Kim, 2018).

Razak et al. (2020) aimed to optimize the placement and sizing of capacitators and implementing conservative voltage reduction (CVR) to minimize power losses and energy consumption, in unbalanced electrical systems, in large-scale office buildings. They used PSOs to determine the optimal solution for capacitor placement and CVR, to minimize total power consumption, and the cost of capacitor installation, while still considering constraints such as voltage magnitude limits. Confirmed by a third-party using the International Performance Measurement and Verification Protocol, a confirmed average of energy savings of 11.48% was achieved (Razak, Othman, Musirin, Yahya and Zakaria, 2020).

Eseye et al. (2019) also addressed the challenges of predicting electricity consumption in buildings, which differ from typical load curves representing large-scale regional or national electricity consumption. Building electricity demand tends to be lower, more volatile, and stochastic, making conventional prediction tools less effective. The study proposes using an integrated approach using three methods, Hilbert-Huang Transform (HHT), a data analysis method used primarily for analyzing nonlinear and nonstationary signals (Michael, Sullivan, Samaratunga and Jha, 2017), PSO, and an adaptive Neuro-Fuzzy system. The purpose was to predict electricity consumption 24 hours in advance, using the area of Espoo, Finland, as a case study, and providing input for smart decision-making in distributed energy management systems for building energy communities. The results demonstrated that the combined approach was able to predict the buildings' electricity needs with lesser error, outperforming previous approaches (Eseye, Lehtonen, Tukia, Uimonen and Millar, 2019).

5.2.8 Building Information Modeling (BIM)

Building Information Modeling (BIM) is a process and technology that enables the creation and management of digital representations of physical and functional characteristics of buildings and infrastructure (Parece, Rato, Resende and Stellacci, 2022).

Alzara et al. (2017) aimed to enhance the efficiency of construction projects by optimizing both project duration and expenses through a plug-in, created by a GA, applied to a five-dimensional BIM interface via the Navisworks software. The research demonstrated that the proposed GA algorithm reduced the project duration by about 20%, while also cutting back on costs. By integrating genetic algorithms with BIM, significant improvements in project cost and cash flows can be achieved, especially for projects of a larger scale (Alzara, Attia, Nahfouz, Yosri and Ehab, 2017).

Conclusion

The bibliometric analysis and comprehensive review of the final 22 studies reveals the pivotal role of AI and SI algorithms in optimizing building energy management across various building energy aspects. The findings highlight the versatility and effectiveness of different AI and SI techniques in addressing energy savings, CO2 concentrations, heating and cooling loads, HVAC systems, and thermal energy storage, as well as in enhancing BIM. For general energy savings in buildings, advancements in predictive algorithms, such as the GOBP model and LSTM networks, have demonstrated significant improvements in accuracy and efficiency.

These models show promise in managing energy consumption more effectively by considering occupant behavior and environmental variables. Regarding CO2 concentrations, adaptive multi-objective optimization algorithms integrated with BPNNs have proven successful in predicting and optimizing indoor air quality and thermal comfort. This approach not only ensures better living conditions but also enhances energy efficiency. Heating load optimization, a critical component of energy efficiency, benefits from hybrid models combining neural networks with optimization algorithms like GAs, PSOs, and ICA. These models provide reliable predictions and management strategies for heating energy requirements, crucial for sustainable smart city planning. Regarding cooling load management, essential for maintaining comfortable interior conditions, there were also improvements using PSOs and GAs. These techniques optimize the operation of cooling systems, resulting in enhanced energy efficiency, and reduced environmental impact. The optimization of HVAC systems, vital for overall building energy management, has been effectively achieved using a variety of swarm-based algorithms, including GOA, WDO, and BBO. These algorithms improve the accuracy of energy load predictions, facilitating better energy management strategies. Thermal Energy Storage (TES), a promising avenue for renewable energy utilization, benefited mostly from ANNs and FL. These methods optimized TES design and operation, ensuring cost savings, and environmental benefits, despite the current limitations of data availability and physical knowledge. For electrical system optimization, enhancement through advanced PSO algorithms showed possibility and potentiality, by addressing challenges in predicting and managing electricity consumption. This approach supports energy savings and efficient decision-making in building energy management. Lastly, the integration of BIM with optimization algorithms like GAs offers significant improvements in construction project efficiency. By reducing project duration and costs, this synergy exemplifies the potential of AI in transforming the building industry.

On the other hand, the bibliometric analysis using VOSviewer validates the current understanding of the field while pointing to new research directions. The densely connected nodes in certain clusters represent well-researched topics, whereas sparsely connected nodes suggest underexplored topics. The temporal analysis in Figure 2 shows a steady increase in publications since 2008, peaking in 2023, indicating growing interest and development in the field. Figure 3 highlights the dominance of engineering, computer science, and energy subjects in literature. Geographically, the research is predominantly conducted in Asia, with China leading the publications. This geographic distribution highlights the global interest but also suggests opportunities for more widespread international collaboration.

The diverse AI and SI algorithms reviewed in this study collectively advance the field of building energy management. They offer innovative solutions for optimizing energy usage, improving indoor comfort, and promoting sustainability in smart buildings and cities. However, there are still very few studies on the potentiality of SI for building energy management, and further research and development is necessary.

References

Abdolrasol, M., Hussain, S., Ustun, T., Sarker, M., Hannan, M., Mohamed, R., . . . and Milad, A. (2021). COA-MLP, which uses COA to optimize the parameters of the MLP, aiming to achieve better accuracy and generalization in the neural network's predictions. *Electronics*, *10*(21).

Albadr, M., Tiun, S., Ad-Dhief, F. and Ayob, M. (2020). Genetic Algorithm Based on Natural Selection Theory for Optimization Problems. *Symmetry*, *12*, 1–31.

Alotaibi, I.M., Abido, M.A., Khalid, M. and Savkin, A. (2020). A Comprehensive Review of Recent Advances in Smart Grids: A Sustainable Future with Renewable Energy Resources. *Energies*, *13*(23), 1269.

Alzara, M., Attia, Y., Nahfouz, S., Yosri, A. and Ehab, A. (2017). Building a genetic algorithm-based and BIM-based 5D time and cost optimization model. *IEE Access*, *99*, 1–1.

Anand, R., Mishra, R. and Khan, R. (2022). Plant diseases detection using artificial intelligence. *In*: M.A. Khan, R. Khan and M. Ansari, *Application of Machine Learning in Agriculture*, 173–90. Cambridge: Academic Press.

Asim, N., Badiei, M., Mohammad, M., Razali, H., Rajabi, A., Haw, L. and Ghazali, M. (2022). Sustainability of Heating, Ventilation, and Air-Conditioning (HVAC) Systems in Buildings: An Overview. *International Journal of Environmental Research and Public Health*, *19*(2), 1016.

Bölek, B., Tutal, O. and Özbaşaran, H. (2023). A systematic review on artificial intelligence applications in architecture. *Journal of Design for Resilience in Architecture and Planning*, *4*, 91–104.

BP Energy Economics. (2023). *Energy Outlook*. London: BP Energy Economics.

Bungau, C., Bungau, T., Prada, I. and Prada, M. (2022). Green Buildings as a Necessity for Sustainable Environment Development: Dilemmas and Challenges. *Sustainability*, *14*(20), 13121.

Chang, K.-H. (2015). Multiobjective Optimization and Advanced Topics. *In*: K.-H. Chang, Computer-Aided Engineering Design, 1105–73. Cambridge: Academic Press.

Chen, D., Zou, F., Lu, R. and Li, S. (2019). Backtracking search optimization algorithm based on knowledge learning. *Information Sciences*, *473*, 202–226.

Dada, A., Joseph, S., Oyewola, D., Fadele, A., Chiroma, H. and Abdulhamid, S. (2022). Application of Grey Wolf Optimization Algorithm: Recent Trends, Issues, and Possible Horizon. *Gazi University Journal of Science*, *35*, 485–504.

Eck, N.J. and Waltman, L. (2010). Software survey: VOSviewer, a computer program for bibliometric mapping. *Scientometrics*, *84*, 523–38.

Elkhatat, A. and Al-Muhtaseb, S. (2023). Combined "Renewable Energy–Thermal Energy Storage (RE–TES)" Systems: A Review. *Energies*, *16*(11).

Emami, S. and Parsa, J. (2020). Comparative evaluation of imperialist competitive algorithm and artificial neural network for estimation of reservoirs storage capacity. *Applied Water Science*, *10*(177).

Eseye, A., Lehtonen, M., Tukia, T., Uimonen, S. and Millar, R. (2019). Short-term Forecasting of Electricity Consumption in Buildings for Efficient and Optimal Distributed Energy Management. *IEEE*, 1103–10.

Esfandi, S., Tayebi, S., Byrne, J., Taminiau, J., Giyahchi, G. and Alavi, S.A. (2024). Smart Cities and Urban Energy Planning: An Advanced Review of Promises and Challenges. *Smart Cities*, *7*(1), 414–44.

Garces-Jimezes, A., Gomez-Pulido, J.-M., Gallego-Salvador, N. and Garcia-Tejedor, A.-J. (2021). Genetic and Swarm Algorithms for Optimizing the Control of Building HVAC Systems Using Real Data: A Comparative Study. *Mathematics*, *9*(18), 2181.

Guerrero-Luis, M., Valdez, F. and Castillo, O. (2021). A Review on the Cuckoo Search Algorithm. *In*: O. Castillo and P. Melin, *Fuzzy Logic Hybrid Extensions of Neural and Optimization Algorithms: Theory and Applications*, 113–24. New York: Springer.

Hafez, F., Sa'di, B., Gamal, S. and Taufiq-Yap, Y. (2023). Energy Efficiency in Sustainable Buildings: A Systematic Review with Taxonomy, Challenges, Motivations, Methodological Aspects, Recommendations, and Pathways for Future Research. *Energy Strategy Reviews*, *45*, 101013.

He, Z., Guo, W. and Zhang, P. (2022). Performance prediction, optimal design, and operational control of thermal energy storage using artificial intelligence methods. *Renewable and Sustainable Energy Reviews*, *156*, 111977.

Hossain, T., Shahid, A., Ali, Y., Saha, S., Jamal, M. and Habib, A. (2023). Fabrications, Classifications, and Environmental Impact of PCM-Incorporated Textiles: Current State and Future Outlook. *ACS Omega*, *8*(48), 45164–176.

IEA. (2019). *Perspectives for the Clean Energy Transition: The Critical Role of Buildings*. Paris: International Energy Agency.

IEA. (2022). *World Energy Outlook*. Paris: IEA.

Iglesias, A., Gálvez, A. and Suárez, P. (2020). Swarm robotics – a case study: Bat robotics. *In*: X.-S. Yang, *Nature-Inspired Computation and Swarm Intelligence*, 273–302. Amsterdam: Academic Press.

Kajjoba, D., Kasedde, H., Olupot, P. and Lwanyaga, J. (2022). Evaluation of thermal comfort and air quality of low-income housing in Kampala City, Uganda. *Energy and Built Environment*, *3*(4), 508–24.

Kumar, A., Chatterjee, J.M., Payal, M. and Rathore, P.S. (2022). Revolutionizing the internet of things with swarm intelligence. *In*: P. Johri, J. Vain, M. Quasim, A. Anand and J. Singh, *System Assurances: Modeling and Management*, 403–436. Amsterdam: Academic Press.

Li, J., Wang, M., Fang, P., Zhao, Q., Zheng, X. and Gao, H. (2024). Modeling and optimization of a passive building HVAC system based on improved PSO and H-J Algorithm. *Energy Reports*, *11*, 2415–38.

Li, L., He, Y., Zhang, H., Fung, J. and Lau, A. (2023). Enhancing IAQ, thermal comfort, and energy efficiency through an adaptive multi-objective particle swarm optimizer-grey wolf optimization algorithm for smart environmental control. *Building and Environment*, *235*, 110235.

Lim, W., Wibowo, A., Desa, M. and Haron, H. (2016). A Biogeography-based Optimization Algorithm Hybridized with Tabu Search for the Quadratic Assignment Problem. *Computational Intelligence and Neuroscience*, *2016*(2), 1–12.

Liu, J., Mei, Y. and Li, X. (2015). An Analysis of the Inertia Weight Parameter for Binary Particle Swarm Optimization. *IEEE Transactions on Evolutionary Computation*, *20*(5).

López, O., López, A. and Crossa, J. (2022). Fundamentals of Artificial Neural Networks and Deep Learning. *In*: O. López, A. López and J. Crossa, *Multivariate Statistical Machine Learning Methods for Genomic Prediction*, 379–425. New York: Springer.

Malik, S. and Kim, D. (2018). Prediction-Learning Algorithm for Efficient Energy Consumption in Smart Buildings Based on Particle Regeneration and Velocity Boost in Particle Swarm Optimization Neural Networks. *Energies*, *11*(5), 1289.

Michael, B., Sullivan, R., Samaratunga, D. and Jha, R. (2017). Elevated Temperature Modal Response and Delamination Detection in Carbon-Epoxy Beams. *SAE International Journal of Materials and Manufacturing*, *10*(1), 34–42.

Moayedi, H. and Mosavi, A. (2021). Double-Target Based Neural Networks in Predicting Energy Consumption in Residential Buildings. *Energies*, *14*, 1331.

Moayedi, H., Canatalay, P., Dehrashid, A., Cifci, M., Salari, M. and Le, B. (2023). Multilayer Perceptron and Their Comparison with Two Nature-Inspired Hybrid Techniques of Biogeography-Based Optimization (BBO) and Backtracking Search Algorithm (BSA) for Assessment of Landslide Susceptibility. *Land*, *12*(1), 242.

Moayedi, H., Nguyen, H. and Foog, L.K. (2021). Nonlinear evolutionary swarm intelligence of grasshopper optimization algorithm and grey wolf optimization for weight adjustment of neural network. *Engineering with Computers*, *37*, 1265–75.

Moayedi, H., Yildizhan, H., Al-Bahrani, M. and Van, B.L. (2023). Appraisal of energy loss reduction in green buildings using large-scale experiments compiled with swarm intelligent solutions. *Sustainable Energy Technologies and Assessments*, 103215..

Nedjah, N., Mourelle, L. and Lizarazu, M. (2022). Swarm Intelligence-based Multi-Objective Optimization Applied to Industrial Cooling Towers for Energy Efficiency. *Sustainability*, *14*(19), 11881.

Ngarambe, J., Yun, G. and Santamouris, M. (2020, March). The use of artificial intelligence (AI) methods in the prediction of thermal comfort in buildings: Energy implications of AI-based thermal comfort controls. *Energy and Buildings*, *211*(1), 109807.

Nguyen, H., Moayedi, H., Jusoh, W. and Sharifi, A. (2020). Proposing a novel predictive technique using M5Rules-PSO model estimating cooling load in energy-efficient building system. *Engineering with Computers*, *36*, 857–66.

Nguyen, H., Zhou, J. and Dou, J. (2019). A comparative study of PSO-ANN, GA-ANN, ICA-ANN, and ABC-ANN in estimating the heating load of buildings' energy efficiency for smart city planning. *Applied Sciences*, *9*(13), 2630.

Ortiz, J., Carrere, J., Salom, J. and Novoa, A. (2023). Energy consumption and indoor environmental quality evaluation of a cooperative housing nZEB in Mediterranean climate. *Building and Environment*, *228*, 109795.

Page, M.J., McKenzie, J., Bossuyt, P., Boutron, I., Hoffman, T., Mulrow, C., . . . and Grimshaw, J. (2021). The PRISMA 2020 statement: An updated guideline for reporting systematic reviews. *BMJ*, *372*(71).

Parece, S., Rato, V., Resende, R. and Stellacci, S. (2022). A methodology to qualitatively select upcycled building materials from urban and industrial waste. *Sustainability*, *14*(6).

Parikh, J.A., DaPonte, J.S. and Vitale, J.N. (1999). Hybrid neural network techniques for storm system identification and tracking. *IJCNN'99. International Joint Conference on Neural Networks. Proceedings*, 4125–28. Washington DC: IEEE.

Razak, M., Othman, M., Musirin, I., Yahya, M. and Zakaria, Z. (2020). Significant Implication of Optimal Capacitor Placement and Sizing for a Sustainable Electrical Operation in a Building. *Sustainability*, *12*(13).

Santamouris, M. and Feng, J. (2018). Recent Progress in Daytime Radiative Cooling: Is It the Air Conditioner of the Future? *Buildings*, *8*(12), 168.

Santamouris, M. and Vasilakopoulou, K. (2021). Present and future energy consumption of buildings: Challenges and opportunities towards decarbonisation. e-*Prime - Advances in Electrical Engineering, Electronics, and Energy*, *1*,100002.

Serbouti, A., Rattal, M., Abdellah, B., Harmouchi, M. and Mouhsen, A. (2018). Application of sensitivity analysis and GenOpt to optimize the energy performance of a building in Morocco. *International Journal of Engineering & Technology*, *7*(4), 2068–74.

Tomar, V., Bansai, M. and Singh, P. (2024). Metaheuristic Algorithms for Optimization: A Brief Review. *Engineering Proceedings*, *59*(1).

Tsilika, K. (2023). Exploring the Contributions to Mathematical Economics: A Bibliometric Analysis using Bibliometrix and VOSviewer. *Mathematics*, *11*(22).

Wu, L., Wu, J. and Wang, T. (2023). Enhancing grasshopper optimization algorithm (GOA) with levy flight for engineering applications. *Sci. Rep.*, *13*(124).

Xiao, W.-S., Li, G.-X. and Tan, L.-P. (2023). A novel chaotic and neighborhood search-based artificial bee colony algorithm for solving optimization problems. *Sci. Rep.*, *13*,20496.

Zadeh, L. (2015). Fuzzy logic: A personal perspective. *Fuzzy Sets and Systems*, *281*(6).

Zhang, C., Ma, L., Han, X. and Zhao, T. (2024). Reconstituted data-driven air conditioning energy consumption prediction system employing occupant-orientated probability model as input and swarm intelligence optimization algorithms. *Energy*, *288*, 129799.

Zhou, Y., Zheng, S. and Zhang, G. (2019). Artificial neural network based multivariable optimization of a hybrid system integrated with phase changing materials, active cooling, and hybrid ventilations. *Energy Conversion and Management*, *197*(6).

Zou, K., Liu, Y., Wang, S., Li, N. and Wu, Y. (2021, December). A Multiobjective Particle Swarm Optimization Algorithm Based on Grid Technique and Multistrategy. *Journal of Mathematics*, *2021*(5), 1–17.

8 Community Energy Balancing

Carlos Coutinho,[1*] Rafael Rodrigues[2] and Manuel Pio Silva[3]

An Intelligent Swarm (IS) comprises a set of multiple intelligent agents (running on or using computers), humans or algorithms performing activities that in some way may contribute to a common objective. Similarly to the behaviour of swarms in nature, a computational IS may include multiple individuals that have the same role or behaviour, but typically it also includes numerous heterogeneous roles and behaviours.

Every agent can independently decide to connect to the swarm (namely by Peer-to-Peer connection), either temporarily or permanently, in order to carry out one or more activities that may involve partial or complete activity flows. The swarm lacks a shared or dominant consciousness and typically lacks a governing body to lead or direct its individuals.

While the traditional application development world follows some well-defined patterns to perform activities, ISs are composed from heterogeneous agents in a chaotic (i.e., non-deterministic) manner. The behaviour of the agents is carefully designed such that their composition shall achieve the emergent behaviour that the programmer had in mind. It is important to note that such a system differs from traditional network programming in that the precise details of the realisation of that goal are unpredictable and often also beyond what we can practically observe—it is comparable to a beehive or an ant colony, where we see the macroscopic result but cannot fully understand every minute action that led to it. Correspondingly, in IS systems the programmer determines the behaviour of individual ants or bees while the purpose is to build the hive or colony.

The development of an IS environment encompasses one or more software projects, each describing one kind of collaboration and its related protocols. It would thus be interesting that the business could define a common environment workspace. Although the workspace itself has no behaviour or (id)entity, it includes one or multiple projects, where each project defines the behaviour and characteristics of one or multiple agents or individuals inside the swarm.

A project usually defines the behaviour and characteristics of some intelligent agents and is commonly described by the definition of states and their interaction in

[1] Caixa Mágica Software, ISTAR, ISCTE-IUL
[2,3] EDP, CNET-Centre New Energy Technologies
Email: rafael.oliveirarodrigues@edp.pt; manuelpio.silva@edp.pt
* Corresponding author: carlos.coutinho@caixamagica.pt

a state machine lifecycle, using annotated graphs of states and labels, defining the events and their behaviour in the state machine. These projects define algorithms that need to consider issues such as security and integrity, but from an individual and contributing perspective.

Considering the deployment of the agents of such system, one or more agents are assigned to each swarm element, their application code is installed on the respective instance (i.e., computer) and configured to match the function of this swarm element in the larger system. In this scenario, a deployed instance can host elements of different swarms, and its user (or the algorithms deployed on it) can interact via a variety of workflows with other parts of the system.

Some of these projects may include properties that are used in other roles as well, meaning that a change in the behaviour or definition of a role may require refactoring actions to be made in other roles as well. It is therefore very useful to define common data types or protocols in reusable software modules.

The TaRDIS (TaRDIS Project, https://doi.org/10.3030/101093006) European Project "Trustworthy and Resilient Decentralised Intelligence for Edge Systems" emerges as a solution aimed at alleviating the complexities of swarm computing and decentralised distributed systems by introducing a novel programming paradigm and providing a comprehensive toolbox to support the development and execution of applications in such environments. As the demand for efficient and correct swarm behaviour grows, there is a critical need for tools that simplify the development process while ensuring the reliability and effectiveness of the deployed systems.

The TaRDIS toolbox targets at boosting the developing framework of swarm systems that exhibit, amongst others, heterogeneous, intelligent, dynamic, and decentralised properties. The developing framework is language-agnostic, leveraging event-driven programming principles, as well as distributed Machine Learning (ML) approaches. In this context, the TaRDIS toolbox can be used by application developers, offering significant abstractions related to the definition of the swarm elements, the local computational resources and datasets, as well as the self-organisation and orchestration capabilities using federated learning, while also putting the focus on the associated communication, security, and data integrity.

The overall goal of the TaRDIS project is to gather end-user functional and non-functional requirements, along with additional technical requirements originating from the development process of specific tools inside the project. However, the project aims much higher than simply to provide a solution to its use-case partners. Its ambition is to be widely adopted, by the industry and related SMEs, but also in academia, as its scope—to provide support to swarm development—is applicable also to the whole software development community. Actually, one of the main targets of the project is precisely the enhancement of the software development experience, to support the developers in the creation of projects using the swarm paradigm.

The project is being evaluated and validated by the definition of a generic pilot use-case, that is capable of demonstrating the swarm paradigm, and one that is able to align with the expectations of a very heterogeneous set of businesses that are interested in developing in this area. Naturally, deriving from this generic use-case, the project also includes strong real industry validation, in multiple and very diverse areas, through the inclusion of four pilot use-cases.

One of these use-case works in the decentralised training and deep learning of smartphones and other smart home elements, through the use of federated learning, towards the objective of achieving the preservation of privacy in these environments. Another use-case is established in the area of distributed aerospace navigation of Low Earth Orbit satellites, promoted by a growing interest in this type of satellite constellation networks, with the target of optimising the costs and effort on the interactions between these satellites and ground stations, by increasing the autonomy of the onboard navigation of these satellites. A third use-case is about optimising the activities in a manufacturing shop floor, promoting the autonomy of the manufacturing robots which perform the various factory activities, and enhancing the collaboration between these shop floor factory elements. The remaining fourth use-case works in the field of Edge Energy management and will be described in more detail in the following pages.

An Energy community is a network of consumers and producers, in a delimited geographical region, who collectively manages and share energy resources.

For electricity, the connection between producers and consumers is established through physical cables with connection points called nodes, establishing an electrical grid. At these grid nodes, voltage levels, frequency and phase need to be always stable within the limits—then, one can say that the grid is balanced for both direct-current (DC) and alternate-current (AC) grids. Having the grid balanced is a sufficient condition to say that production meets consumption. Nowadays, the integration of Renewables, without primary energy costs associated, allows for a reverse approach. For example, one can defer energy consumption to align with solar daily peak production, a concept commonly known as demand-side flexibility. This flexibility can be achieved through the use of adaptable loads, such as Electrical Vehicles (EV), and the creation of demand profiles.

Geography and demographics drive electric grid sizing. Technically, cables and nodes should support the power flow but as distance from generation sources increases, losses in the cables become non-negligible. So, having local production nearby demand is far more efficient than energy transportation. In this context, Energy communities play an increasingly important role as demonstrated in the work done in (Cardoso et al., 2023).

Although there are several Energy communities already in place, centralisation on the role of the community aggregator is a limitation to citizens' active participation in energy trading, opting for best value for the money, according to its own criteria.

If each peer is able to connect within its community and make automatic agreements all the processes would be optimized, it could run 24h/7d with reduced human intervention and costs would be lessened also. This is the perfect ground for applying the TaRDIS toolbox.

The interactions between peers (shown in Figure 1) within a community take the primary part of the specifications, while the format of the message stands as the second most important. The former will be described using communication diagrams, depicting the interaction between prosumers in both the roles of a producer and consumer (producer + consumer = prosumer), and the community Community Orchestrator (CO) responsible for the external interactions with other communities' orchestrators and with the DSO. The latter is statically described, remaining simple and stable, and will be used in different contexts, with different purposes, namely for information exchange.

All messages have the format described in Table 1.

Table 1 Format of the messages.

Time period: hh:mm – hh:mm	**Time period** is the time slot considered for the energy transaction.
Energy (kWh): real value	**Energy** is the energy that is being negotiated/acknowledge/missing/surplus in the time period mentioned.
Price (optional)	**Price** is an optional parameter that can be or not included in the negotiation model.
Acknowledge: Boolean	**Acknowledge:** **True** = accepted/concluded; **False** = refused/aborted/failed/requested.

With this foundational understanding, we can now outline the key objectives that drive the efficient and sustainable operation of the energy community. These objectives, initially described in (Seco et al., 2024), are here described by precedence:

1. Maximise the use of Renewable energy inside an Energy community.
2. Guarantee that all consumers in the community have energy supply.
3. Reduce the number of messages between peers within the communication network.
4. Use the same application (code) for all actors.
5. Have a system that can, in the future, incorporate a local energy market with hourly bids.

For the system to operate, two stages are needed: the first stage is the agreement on supply-consume planning—ex-ante working mode refers to planning the conditions for trading in the next hour and second stage is the running mode.

Demand should lead the process as human needs are based on energy consumption. Demand side management is not yet considered in the proposed scenarios, just because it would increase the number of scenarios without replication potential.

As in a market, a match occurs when the demand request and supply offer are met. In this case, an energy request from each consumer triggers bids from energy

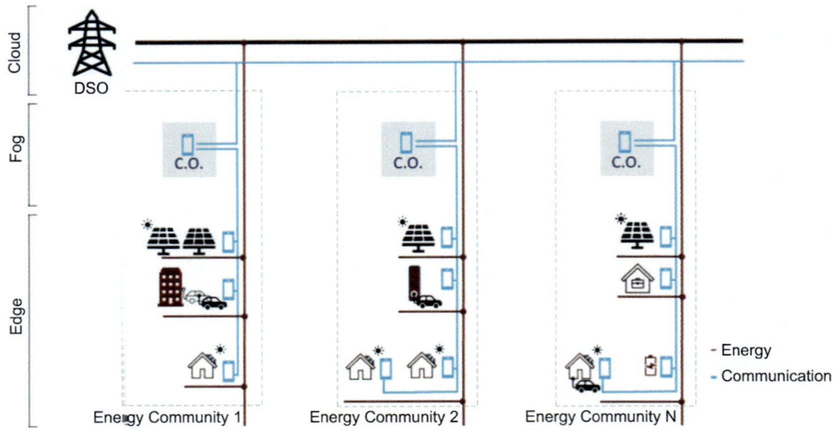

Fig. 1 Energy use case overview.

suppliers and each consumer can choose its suppliers, for the period, based on its own criteria (price, source, energy peak, etc.).

For the energy use case the following assumptions were taken:

1. The Grid Distribution System Operator (DSO) is the highest order energy provider. The grid should be considered as the "option of last resort" when energy consumption/production cannot be successfully balanced within communities. Consumption from the grid operator is more expensive than from neighbour communities, because surplus will have the lowest price.
2. The Energy Community is a digital smart grid (for instance a set of households, buildings, neighbourhood, or a city) consisting of several energy producers and consumers that can exchange energy among themselves. Additionally, energy communities can also request supply or provide energy to neighbouring energy communities. All operations are acknowledged by the Community Orchestrator (C.O.)
3. Prosumer represents a community member (i.e., a household or an EV charger) that is registered for consuming or producing energy. Producing energy in households currently assumes some renewable energy source such as photovoltaics (PV), micro wind turbine generators, but can even come from Battery Energy Storage Systems (BESS).

The basic prerequisites for the energy use case are the following:

1. An adequate communication link between the DSO-CO-Consumer.
2. A physical electrical connection between DSO-CO-Consumer.
3. Availability of end users participating in energy exchange market.

Six scenarios were defined (as shown in Figure 2), with two major groups: 'Plan' with four scenarios and "Run" with two scenarios.

The initial two scenarios in the planning phase are common in every one-hour-time period. They encompass the planning of generation and consumption. The last two planning scenarios address specific situations in the energy community when the energy is not balanced. The first situation involves a deficit of energy, while the second deals with a surplus. The final two scenarios describe the "<<RUN>>" mode, where we have already entered into operation, using the prepared working conditions. In this mode, we can either have normal operation or operate with faulty conditions.

In the following sequence diagrams, we describe the six scenarios for this use case, considering always one-hour periods. The scenarios are divided into two groups, one regarding the planning work 'with' or simply "<<PLAN>>", represented under the left side of Figure 2: this is the place where all peers agree on Energy production, distribution, and consumption just before the new period starts. Another group is where real-time operation is described as "<<RUN>>" on the right side of Figure 2. In summary, the figure displays two stages of the energy UC, in the leaf nodes it is possible to observe the six scenarios, for each of them below was produced a sequence diagram with the goal of specifying the base dependency and minimum configuration for functionality.

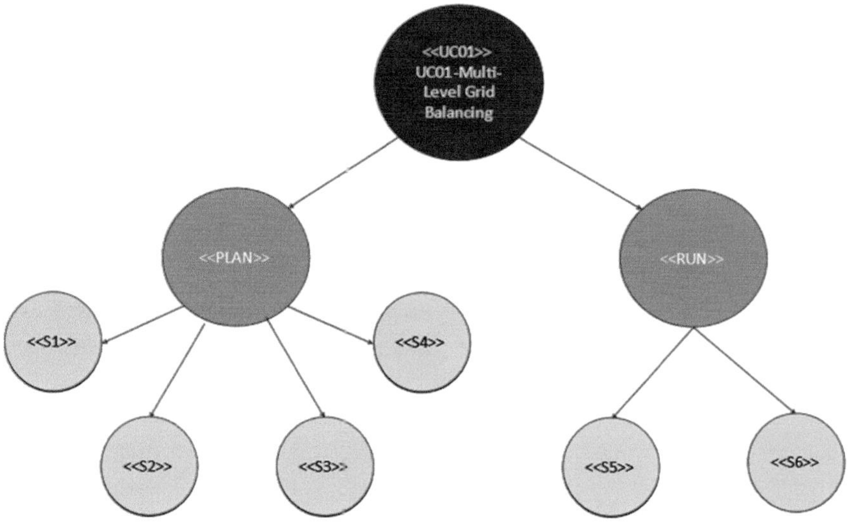

Fig. 2 Domain overview (Rodrigues et al., 2024).

For context and enhanced readability in the upcoming scenarios, the involved actors are briefly introduced in Table 2.

Their interaction is visually represented in the following Role diagram, providing a more detailed insight into their dynamic relationships and roles, as shown in Figure 3.

Table 2 Energy Actors.

Actors	
Name	*Description*
Consumer (User)	Citizen who is end-user of electricity (e.g., EV, home appliance, house or building).
Prosumer	Consumer that can also generate energy using a Distributed Energy Resource (e.g., PV, Residential Battery or Battery from an EV capable of Vehicle to Grid/Home V2G or V2H).
Distribution System Operator (DSO)	Company responsible for the grid operation, distributing, and managing energy from the generation sources to the final consumers (e.g., EDP).
Community Orchestrator (CO)	Entity responsible for communicating with all actors ensuring the good planning and running of the energy community.

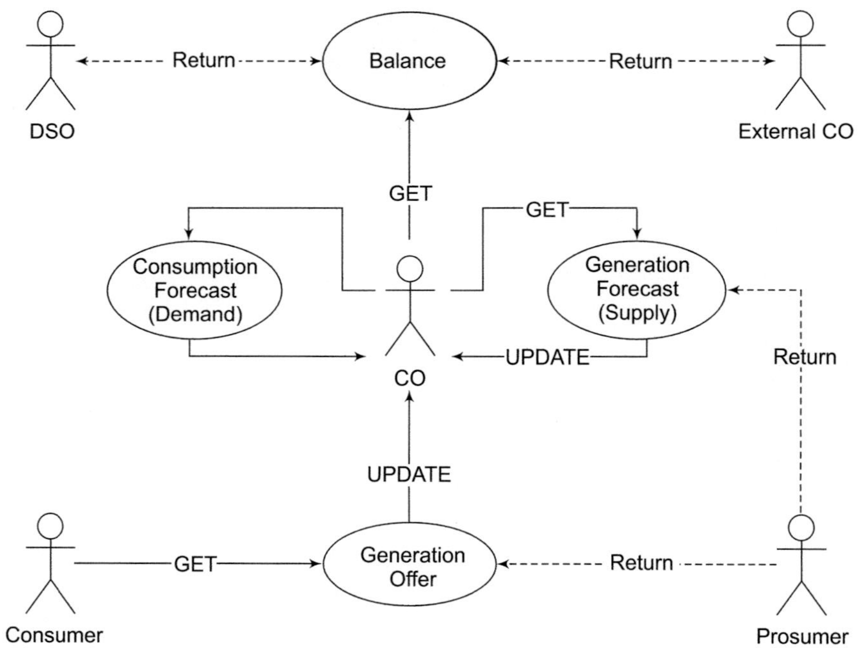

Fig. 3 Actors Role diagram (Rodrigues et al., 2024).

1. Scenario 1: Ex-ante working Energy generation forecast for the next hour

The first scenario outlines the process for obtaining the energy generation forecast for the next hour. It begins with the community orchestrator requesting the expected generation from the producers within the community for this time slot. The process

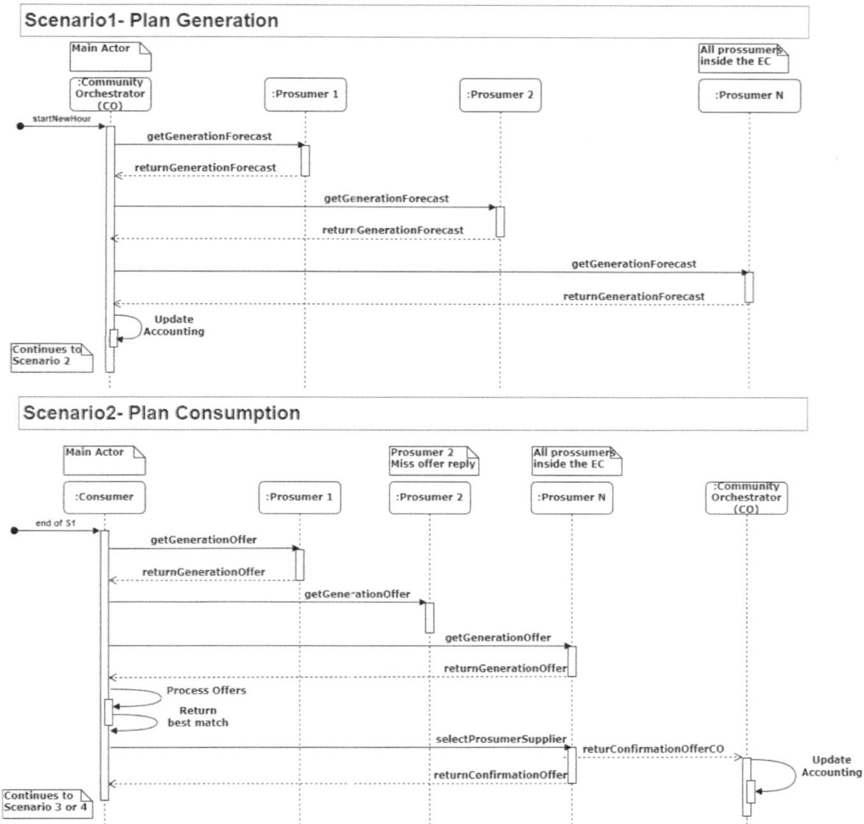

Fig. 4 Scenarios 1 and 2 sequence diagrams.

then moves to the collection of feedback, concluding with the update of the Community Orchestrator's records of generation. The scenario describes the stage when the Community Orchestrator requests the producers how much energy will they produce in the next (one hour) period calculating the total available, as can be seen in Figure 4.

2. **Scenario 2: Ex-ante working Energy consumption forecast for the next hour**

The second scenario outlines the process of acquiring the energy consumption forecast for the next hour. The process initiates from the consumer side, where a consumer is unable to meet their own energy requirements for this timeslot, begins by requesting available energy from his prosumer peers. Subsequently, the consumer seeks energy offers from peers (producers), proceeding to the third step of consumer selection of the best offer. The process concludes with the communication of information to the selected producer and the community orchestrator for the updating of accounting records, as seen in Figure 4.

Consumers take Scenario 1 as a tacit signal from the orchestrator that the new period is about to start and broadcast their consumption. Then the negotiation between producers and consumers starts and, using its own criteria, each consumer books production from producers. The model may or may not take price into account. The Community Orchestrator can now totalise the community's internal consumption.

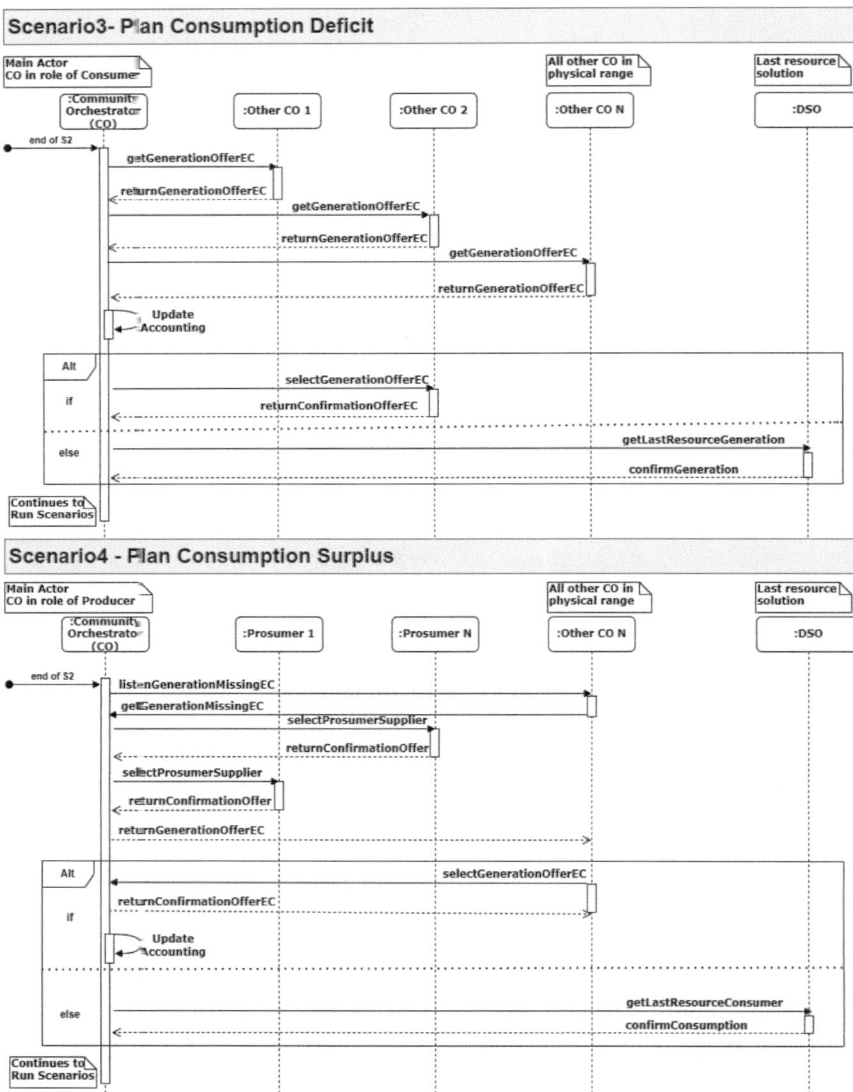

Fig. 5 Scenarios 3 and 4 sequence diagrams.

3. **Scenario 3: Ex-ante working with total Energy consumption forecasted for the next hour and balance of deficit**

In this scenario, the community is unbalanced after operations in Scenarios 1 and 2, requiring energy from external sources. The community orchestrator takes on the role of a consumer and requests energy from other community orchestrators. Two possible situations arise from this: either the remaining energy needs are fulfilled by other community orchestrators, or if not, the community orchestrator requests the remaining deficit to the DSO, as seen in Figure 5 top part.

4. **Scenario 4: Ex-ante working with total Energy consumption forecasted for next hour and balance of surplus**

Here the perspective is complementary to Scenario 3, as seen Figure 5 bottom part, where the community, after operations in Scenarios 1 and 2, is unbalanced with a surplus of energy. The community orchestrator takes on the role of a producer and offers energy to other requesting community orchestrators from Scenario 3. Two possible situations arise from this: either the surplus of energy is fulfilled by the other requesting community orchestrators, or if not, the community orchestrator injects the remaining surplus in the grid managed by the DSO.

Scenario 5: Running in Normal operation mode

In the normal operation mode, following Scenario 2, the consumer receives the energy offer from the selected producer, acknowledges this offer, informs the community orchestrator, and asks the producer to initiate the energy transaction. Finally, the consumer notifies both the producer and orchestrator of the total energy consumed, allowing them to verify and update the records (Figure 6).

6. **Scenario 6: Running with faults**

In this final scenario, as seen, Figure 6, two types of faults may occur, either from the supply side or the demand side. In the case of a supply fault, the consumer notifies the community orchestrator of the issue. The orchestrator resolves the energy problem using information from Scenario 1. If it progresses to Scenario 3, the grid will always fulfil the consumer's needs. The energy would be inject into the grid, and consumed by the DSO in other point, where the producer failed to deliver energy, if everything goes well energy will always close the loop. If maintenance is required, it will be carried out by a human, and if the fault is recurring, the producer may be excluded from the pool. In the case of a demand fault, the producer notifies the community orchestrator and ceases the production. If unable to cease production, the producer starts injecting into the grid and the consumer does not need to be alerted.

Table 3 shows all six scenarios and their respective interactions.

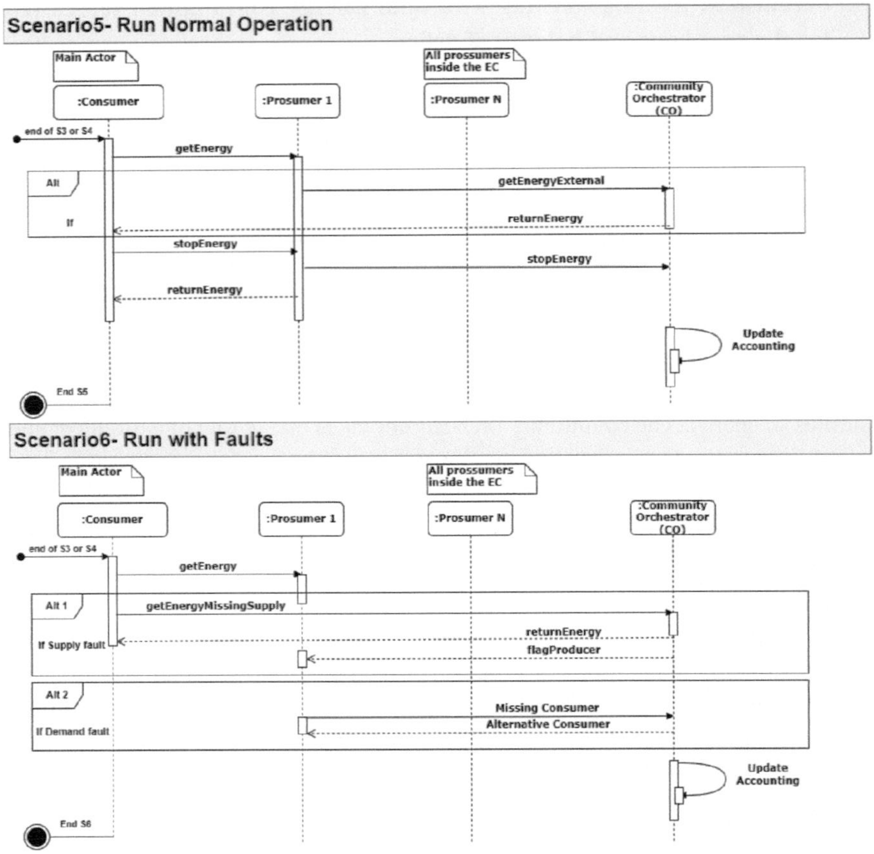

Fig. 6 Scenarios 5 and 6 sequence diagrams.

From Table 3, it is possible to generate the state machine that will run for each time period, shown in Figure 7.

Table 3 Energy scenario conditions.

Scenario Conditions						
S. No.	*Scenario Name*	*Scenario Description*	*Main Actor*	*Trigger Event*	*Pre-condition*	*Post-condition*
1	Plan-Generation	Plan Generation in community next hour.	CO	Start (New hour)	NA	S2
2	Plan-Consumption	Plan Consumption in the community next hour.	Consumer	S1	S1	S3 S4 S5

Contd.

Table 3 *Contd.*

Scenario Conditions						
S. No.	*Scenario Name*	*Scenario Description*	*Main Actor*	*Trigger Event*	*Pre-condition*	*Post-condition*
3	Plan-Deficit	Energy deficit in the community next hour.	CO		Balanced-No Missing energy-Yes	S5 S6
4	Plan-Surplus	Energy surplus in the community next hour.	Consumer	S2	Balanced-No Missing energy-No	S5 S6
5	Run-OK	Normal operation all running ok.	Consumer	S2 S3 S4	Run Faults-No	End
6	Run-Fault	Operation fault, supply or demand fault.	Consumer	S2 S3 S4	Run Faults-Yes	End

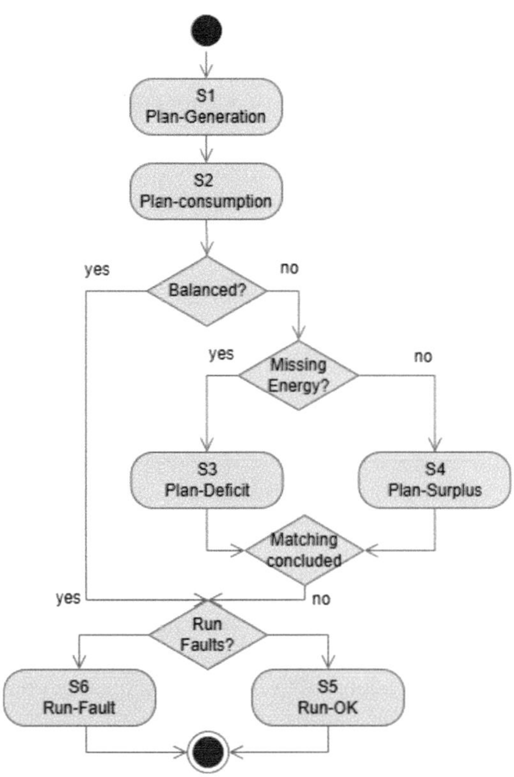

Fig. 7 Energy State Machine.

In its current development status, the TaRDIS toolkit comprises over 40 tools, each addressing various potential applications. Not all tools will be used in the specific Energy use case, but only a subset of the overall toolbox, which is precisely the purpose of having a mashup of tools that have multiple different ways to be combined so to achieve the best support for the swarm development of each case.

The Multi-level Grid Balancing use case will be deployed as a set of three layers, as shown in Figure 1, where the fog layer will be implemened by a swarm of CO elements and the edge layer will be implemented by a swarm of edge devices. In this sense, TaRDIS will contribute with tools to support the development the applications that will populate these swarms, such as:

The Dynamic Condition Response (DCR) choreography editor: This tool aims to ease the writing of applications that model the behaviour of each actor in the Energy Community. The used language includes the capability of querying the state of other input events or other data elements (computation events). A connection to the Machine Learning (ML) APIs is essential to guide the matchmaking process in this use case. The semantics of query expressions used in the DCR graphs can be powered by ML mechanisms and APIs. External applications can register to execute events of the workflow having the appropriate emitter role. Such applications can also register in the system, to listen to events that they are set to receive and that they depend on to proceed. The enableness of an event, which declares that the workflow is in a state that allows the execution of an event, should be established locally based on the notice of execution of events in the system, propagated/replicated through the decentralized system (swarm).

The IFChannel (Information Flow Channel) tool is targeted to assure privacy and security of confidential information exchanged between peers (e.g., when producer and consumer agree on details like the supply's time range or price). This is an important feature as having this information accidentally leaked may distort local market operation and grid stability.

The ReGraDa-IFC (Information Flow Control) and DCR Choreographies tool is a compiler and type checker, helping the tailored use case code and flows to fit the requirements for information security in DCR graphs that use cryptographic channels over public networks (e.g., at community orchestrators' swarm level which may be geographically apart).

Federated Learning (FL) training will be performed by the Fedra tool. This tool will be used to host the decentralized federated learning of two types of ML models that are deployed locally on several edge devices/nodes (clients). The two ML models that have been developed and will be demonstrated in the multi-level grid balancing use case involve:

(i) the prediction of the future energy needs and the prediction of the future energy generation from renewables of a smart home (which is considered the edge node) using Long Short Term Memory (LSTM) models. These models will be deployed at each edge node (2-3 smart homes will be considered during the demonstration) and start the training process with the local datasets;

(ii) Deep Reinforcement Learning (RL) models for optimization of the energy cost related to the power consumed in each smart home for temperature comfort. In this context, the Deep RL model targets to regulate the input power of a Heating, Ventilation and Air Conditioning system (HVAC), keeping the temperature between the comfort limits, while at the same time minimizing the energy exchange between the smart home and the grid, thus, minimizing the energy cost and promoting the consumption of local renewable energy. Although this Deep RL model is deployed at a single node (smart home), the target is to also demonstrate federated learning among several homes, by using either Flower or Fedra. The idea is to showcase the Deep Federated RL in a centralized federated learning framework, where the clients are smart homes inside an energy community and the server is the community orchestration.

The Pruning tool (lightweight models) aims to provide more lightweight ML models at the edge devices (smart homes), without significant degradation of their accuracy. To this end, the pruning tool will consider the pre-trained LSTM models that are trained for power generation and power consumption. The updated lightweight version of the LSTM models will be trained, tested and its performance will be quantified (accuracy rate compared to the original model, memory and CPU requirements, inference latency).

The Membership service will allow an element to connect to all other elements in the swarm(s) for which it has appropriate credentials (e.g., edge swarm elements can connect with each other and to the community orchestrator but not directly to other elements from another energy community).

The Babel framework is implementing the core functions for connecting the swarm elements, by implementing the overlay network, connecting every single swarm member, and enable higher level developments. In this way each edge swarm and community operators' swarm can share the same communication definitions.

To summarize this chapter, the introduction of intelligent swarm systems through TaRDIS is anticipated to bring several benefits. TaRDIS outcomes will include consumption and generation forecast as a way of short-term grid balance planning, data layer in the communication model within the community to coordinate offers and needs and an abstract way of programming all of this. The benefits for the Energy sector include enhanced capabilities for establishing new energy communities, improved grid operations, and increased observability resilience due to the finer granularity in measurement and control. Another relevant outcome is the potential maximisation of locally produced energy, which can have a positive environmental impact given its renewable nature and reduced distribution losses. Further improvements/benefits are being discussed and aligned with the researchers from academia and IT partners. To assess these metrics, a set of KPIs are described in (Rodrigues et al., 2024). It is possible to estimate the CO2 eq. avoided by the use of Renewables within a community. Lastly, for the setup of new communities, this verification will be carried out through the utilisation of a TaRDIS developed application designed to execute this task.

Acknowledgement

The authors wish to thank the support from the TaRDIS Project. This project is funded by the European Union (TaRDIS, 101093006). Views and opinions expressed are however those of the author(s) only and do not necessarily reflect those of the European Union. Neither the European Union nor the granting authority can be held responsible for them.

References

Cardoso, C., Rodrigues, R.O., Gomes, F.C., Piedade, J.L., Mendes, G., Figueiredo, R., . . . and Cabral, D. (2023). eNeuron project: Facilitating the energy transition in a military campus by optimising a local energy community. *In*: *27th International Conference on Electricity Distribution* (*CIRED 2023*), 895–99. Rome, Italy. doi:10.1049/icp.

Matias, S., Mateus, J., Pereira, M., Silva, T., Furtado, A., Ziras, C., . . . and Morais, H. (2023). *V2X Integration in Self-Consumption Energy Management System*. doi:10.1049/icp.2023.0437.

Rodrigues, R.O., Silva, M.P., Kuhn, R., Tsigkari, D., Scalas, A., Coutinho, C., . . . and Vázquez, D. (2024). *TaRDIS D2.2: Report on Overall Requirements Analysis*. Retrieved from https://www.project-tardis.eu/wp-content/uploads/sites/101/2024/02/D2.2-V1.1-Final.pdf.

Seco, J.C., Soares, C., Ferreira, C., Leitão, J., Ravara, A., Kourtellis, N., . . . and Silva, M.P. (2024). *TaRDIS D3.1: First Report on Programming*. Retrieved from https://www.project-tardis.eu/wp-content/uploads/sites/101/2024/02/TaRDIS_D3.1-final.pdf.

TaRDIS Project. (n.d.). Retrieved from https://www.project-tardis.eu.

9 Design for Healthcare in Current and Future Cities

Exploring Assessment Tools and Methods for Evaluating Sustainability and Inclusion in Healthcare Complex Infrastructures

Andrea Brambilla,[1] Erica Isa Mosca[2*], Tianzhi Sun[3] and Stefano Capolongo[4]

1. Introduction

1.1 Healthcare Environments

The quality of healthcare environment significantly influences the functionality of healthcare operations, the well-being of patients, and the efficiency of healthcare providers. Therefore, to achieve the goal of environmental and social sustainability, providing high-quality and patient-centered care, it is essential to evaluate the quality of healthcare environments in international contexts (Cambra-Rufino et al., 2021; Ulrich et al., 2010).

However, despite the need to have innovative and flexible hospitals in recent times, 60% of them in Italy are more than 50 years old, which is the optimal lifespan estimate (Astley et al., 2015; Lega et al., 2011). Additionally, 50% of the time hospitals cannot host the current organizational models (Brambilla and Capolongo, 2019). Several reports highlight this low quality and low investment in the built environment of hospitals (Ricciardi, 2015; Sdino et al., 2021). Facing the great challenges, there are fundings like the National Recovery and Resilience Plan (NRRP) in Italy as a part of Next Generation EU (NGEU) (PNRR, 2021), and the UK Health Infrastructure Plan (HIP) (Department of Health and Social Care, 2020), aiming at improving and supporting the renovation of the healthcare system. Therefore, new assessment strategies are needed to evaluate the quality of such investments in complex healthcare infrastructures for future cities and societies (Jaušovec and Gabrovec, 2023). The question concerns how to assess those investments, encompassing their physical and organizational qualities, and how to analyze the data collected within hospitals.

[1,2,3,4] Design & Health Lab, Department of Architecture, Built Environment, and Construction Engineering, Politecnico di Milano
Email: andrea1.brambilla@polimi.it; tianzhi.sun@polimi.it; stefano.capolongo@polimi.it
* Corresponding author: ericaisa.mosca@polimi.it

1.2 Why Focusing on Environmental and Social Sustainability?

From an environmental perspective, implementing energy-efficient systems, sustainable waste management practices, and eco-friendly materials can significantly reduce the environmental impact of healthcare operations. Sustainable healthcare environments also promote better public health by reducing pollution and contributing to a cleaner and safer environment. Moreover, sustainable healthcare environments can lead to significant cost savings in the long run. In addition to lowering operating costs through energy-efficient design, well-designed facilities can also enhance the lifespan of buildings and reduce the need for frequent renovations, leading to more efficient capital allocation.

Social sustainability in healthcare environments focuses on creating inclusive spaces, promoting well-being, and supporting the diverse needs of patients, staff, and the community (Figueiro, 2008) as the name implies, a movement to ensure long-term, efficient utilization of resources. In fact, sustainability does not imply that resources should be restricted or subsidized. (McDonough and Braungart 2002). Evaluating and ensuring accessibility, comfort, and safety in healthcare settings contribute to a more equitable health service delivery (Mosca et al., 2022). Furthermore, sustainable healthcare environments can serve as models for the community, promoting health and wellness and fostering a culture of sustainability (Buffoli et al., 2015; Capolongo et al., 2016).

1.3 Evaluation of Healthcare Environment

Rigorous research and systematic evaluation, serving as repositories for the highest quality information available, constitute the cornerstone of healthcare facility design (Ulrich et al., 2008). This foundational principle underscores the importance of leveraging empirical evidence and comprehensive analyses to inform the architectural and operational planning of healthcare environments (Haapio, 2012). Such an evidence-informed approach ensures that design decisions are not only based on theoretical principles but are also grounded in verified data, reflecting best practices and innovations in the field. Consequently, the integration of solid research findings and detailed evaluations into the planning process is instrumental in optimizing the functionality, safety, and user experience of healthcare facilities, thereby enhancing patient care and staff efficiency.

The assessment and enhancement of healthcare environments from a sustainability perspective are paramount (Buffoli et al., 2015). It not only ensures the responsible use of resources and the reduction of environmental impact but also promotes economic efficiency and social well-being. Sustainable healthcare environments are a critical step toward a holistic approach to health and well-being, aligning with the broader goals of Sustainable Development (Brambilla and Capolongo, 2019; Dolcini et al., 2023).

2. Methodology

This article presents an in-depth examination of two evaluation instruments designed for the healthcare environment, with a specific focus on aspects of environmental and social sustainability. Through a detailed exploration of these tools, the article delves into the significance of the built environment's quality within the healthcare and public health sectors. It underscores how the quality of physical spaces directly influences health outcomes and the well-being of both patients and healthcare providers.

This approach has been specifically explored within the didactic module in Architectural Technology "Building Sustainability and Design through Digital Tools" of the ATHENA Erasmus + International Project "University goes Digital for a Global Sustainable Education" held at ISCTE, University Institute of Lisbon with Politecnico di Milano, Department ABC contribution (https://athenadigitaluniversity.eu).

Specifically, this article outlines the methodological framework employed in developing these evaluation tools, emphasizing the necessity for methodical research and precise quantification in the study of the built environment. This approach highlights the importance of evidence-based strategies in enhancing the design and functionality of healthcare facilities, ultimately aiming to improve inclusive patient care, staff efficiency, and the overall sustainability of healthcare services.

2.1 Methodological Development of SustHealth Version 2

SustHealth Version 2 (V2) was designed as an evidence-informed assessment tool to provide decision support for hospital strategic management. Its evaluation focuses on the renovation and implementation of operative facilities, specifically from the perspective of architecture and the built environment. The tool was developed, weighted, and tested to ensure its effectiveness in guiding strategic decisions in these areas.

The tool emphasizes the crucial role of the multiple criteria assessment tool in fostering improvements in hospital infrastructure. The tool's comprehensive approach includes evaluating the physical condition of buildings, the efficiency of internal processes, and the impact on staff and patient well-being. This holistic perspective is pivotal in identifying areas that require attention and guiding targeted interventions. As shown in Figure 1, the methodology of SustHealth V2 included tool development, weight assignment, and tool test.

Tool Development: The methodological framework of this research was critical, beginning with an evaluation of the current state-of-the-art tools and a comprehensive literature review on evidence-based design, quality improvement, and hospital management. Then a critical assessment of the existing tools through SWOT analysis identified its strengths, weaknesses, opportunities, and threats.

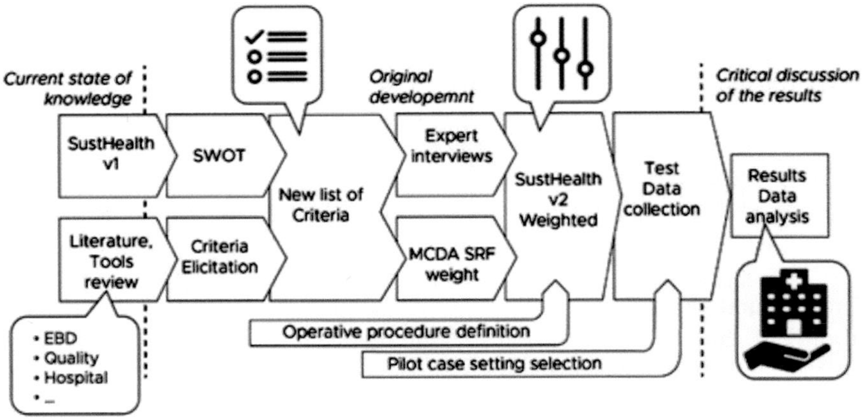

Fig. 1 Methodology of SustHealth v2.
Source: Authors.

Criteria elicitation incorporated scientific literature, international best practices, and feedback from focus groups involving students, experts, and practitioners.

Through a SWOT analysis, existing tools were scrutinized to elicit criteria from the literature, reviews, and current knowledge, leading to the creation of a new list of criteria combining both existing and new elements. The resulting framework, starting from the original and undergoing structural implementation, simplification, optimization, and updates, led to the creation of the new tool, SustHealth V2, showcasing a systematic approach to enhancing healthcare facilities' evaluation.

Weight Assignment: The list of tool criteria experienced validation through expert interviews, semi-structured interviews, and focus groups, subsequently being weighted using a multi-criteria decision support system methodology known as the "MCDA SRF DCM" method or Simon Roy Figueras (SRF) procedure (Corrente et al., 2019; Figueira and Roy, 2002). This innovative approach facilitated the development of a tool equipped with a weighting system and an operational procedure for data collection and analysis.

Experts recognized nationally or internationally for their contributions to publications, projects, and their professional network, played a crucial role. The research, conducted in Italy and Sweden, incorporated experts affiliated with various institutions, each specializing in different aspects of healthcare facilities, including management, social, and environmental spheres. The MCDA SRF DCM method allowed these experts to visually rank the importance of criteria and indicators, further refining the tool's precision through a competition phase utilizing DecSpace web platform. This phase translated visual rankings into quantifiable data for comparison and refinement as described in previous studies (Brambilla et al., 2021).

Tool Test: The tool's efficacy was tested through pilot case studies, involving data collection, analysis, and a critical discussion of the findings. The testing methodology encompassed the input, activity, and output phases. Hospitals provided technical documentation for the input phase, followed by documentation analysis, surveys, site visits, and interviews during the activity phase. The output phase includes the delivery of a final report, improvement actions, and simulations. Despite its complexity, the process was expedited by a binary system (yes/no checklist), ensuring speed, effectiveness, and reliability. The tool was applied to various hospital areas, comparing public and private hospitals to assess its versatility and effectiveness in the international context and opened up opportunities for market implementation (Brambilla et al., 2022).

2.2 Methodological Development of Design for All A.U.D.I.T. (Assessment Usability Design and Inclusion Tool)

Design for All A.U.D.I.T. (Assessment Usability Design and Inclusion Tool) is a tool that assesses accessibility and inclusion according to Design for All (DfA) principles, through a framework of informed-based scientific indicators (Mosca and Capolongo, 2023). DfA is the design approach for human diversity, equality, and social inclusion (European Institute for Design and Disability, 2004), it strives for universal accessibility regardless of age, gender, or ability, endorsed by European standards (e.g., EN17210:2021) for designing built environment through a performance-based approach for a wide range of people (Mosca et al., 2019; Mosca and Capolongo, 2018). DfA A.U.D.I.T. evaluates the level of DfA with qualitative and quantitative criteria, to support decision-makers to evaluate space inclusivity for planning and renovation guidance.

DfA A.U.D.I.T. adopts a rigorous and scientific approach, emphasizing the detailed analysis of all groups' needs and objectives. The development methodology involves end users at each step as the association of people with disabilities and experts in accessibility and inclusions. This approach not only enhances the relevance and usability of the building but also aligns with the principles of person-centered design, promoting solutions that are universally accessible and beneficial and therefore contribute to social sustainability.

Tool Development: The process of developing the tool has been structured in three phases 1. Analysis; 2. Elaboration; 3. Application-Validation as shown in Figure 2. Various methods were employed to identify categories, criteria, and indicators for the development of the DfA evaluation framework. Initially, a systematic literature review was conducted to gather existing knowledge on DfA as it relates to performance assessment in the built environment. This analysis identified three key qualities fundamental to DfA evaluation: Physical, Sensory and Cognitive, and Social. The development process also included six workshops and focus groups involving users with impairments and experts, where cognitive mapping analysis was utilized. This approach facilitated the definition of objectives for evaluating spaces according to each criterion identified from the literature review. The examination of four different hospitals provided additional insights into the criteria and indicators relevant to the healthcare environment.

Tool Validation: The framework was validated through interviews with national and international experts, ensuring its applicability and relevance across various contexts. Then, the tool was first demonstrated in a private hospital of Buffalo (U.S.A.) where objective data was gathered based on the tool's indicators and criteria across different building areas. A concurrent questionnaire correlated the tool's objective assessments with subjective user experiences, employing similar evaluation items. Additionally, the tool underwent review by experts at the University at Buffalo's IDEA Center (White and Mosca, 2022). The final validation occurred in an Italian hospital in Milan.

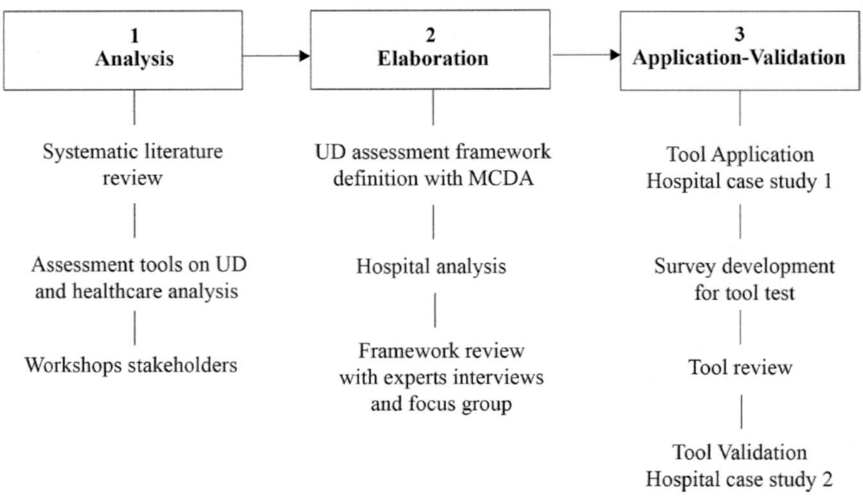

Fig. 2 Methodology of Design for All A.U.D.I.T.
Source: Authors.

3. Result

3.1 The Structure and Content of SustHealth V2

Tool Framework: This framework is intricately structured, comprising macro areas, criteria, indicators, and variables, which collectively form the operational methodology as shown in Figure 3. Each indicator is assigned a unique value, contributing to the criteria, which in turn aggregate to define macro criteria. Specifically, the framework encapsulates social, environmental, and organizational qualities as its cornerstone macro criteria, with Figure 3 illustrating the composition of these areas within the tool's framework.

The social qualities domain encompasses indicators related to non-clinical social impact activities within the hospital, aiming to enhance the well-being, satisfaction, safety, and inclusion of all users. This area emphasizes: 1.1 Sustainable accessibility, 1.2 Security and control, 1.3 Empowerment, 1.4 Inclusion and attractiveness, 1.5 Health promotion, and 1.6 Visual environment. Environmental qualities focus on the hospital's impact on the environment, addressing climate change and Sustainable Development Goals (SDGs), including 2.1 Policy and education, 2.2 Waste management, 2.3 Energy Management, 2.4 Comfort and perceived well-being, 2.6 Wayfinding and Economy. Organizational qualities cover aspects critical to the hospital's overall functionality, emphasizing 3.1 Patient safety, 3.2 Certifications, 3.3 Future proofing facility, 3.4 Logistics and efficiency, 3.5 Technological innovation, and 3.6 Facility management.

Weight: The weighting process assigns significance to each macro area and its respective criteria, resulting in a tool that facilitates both a visual assessment and the derivation of minimum, maximum, or average scores based on hospital performance during the checklist application. The assessment encompasses sheets detailing criteria, indicators, their impact, and measurable variables, enabling quantification through a simplified binary checklist and subsequent score compilation. Semi-structured interviews and the application of the Simon Roy Figueras (SRF) procedure for the elicitation of weights criteria through the Deck of Card Methods allowed us to involve international experts and decision-makers from the healthcare sector supporting the weight definition. The process and results of the weighting are in detail explained in the previous research (Brambilla et al., 2020).

Tool Testing: Tool testing involves a weighted sum calculation of standardized scores collected during the evaluation phase, with data analysis enabling the examination of relationships between indicators and criteria. This analysis offers insights into areas such as social sustainability and quality, where, for example, sustainable accessibility improvements can significantly impact hospital organizations, irrespective of their proximity to urban centers. The environmental analysis highlights the importance of direct management in energy and waste, whereas organizational aspects underscore patient safety and the strategic implementation of future-proofing measures for flexibility and reference project development. Two Italian hospitals are analyzed in the Pilot test.

After the validation of the tool, SustHealth V2 was used in practical research testing 14 German hospitals for their Quality and Sustainability (Brambilla et al., 2020). The 199 variables have been translated into an online survey in German, and the survey was created on the online survey platform and diffused via email. Social Quality macro area scored a mean of 0.65 and a median of 0.68. Environmental Quality macro area scores a slightly lower mean and median with 0.51 and 0.49. Organizational Quality scores similarly to Environmental Quality, with a mean of 0.50 and a median of 0.49. The scores of each category demonstrate the hospitals' strengths and weaknesses in terms of quality and sustainability, and the evaluation

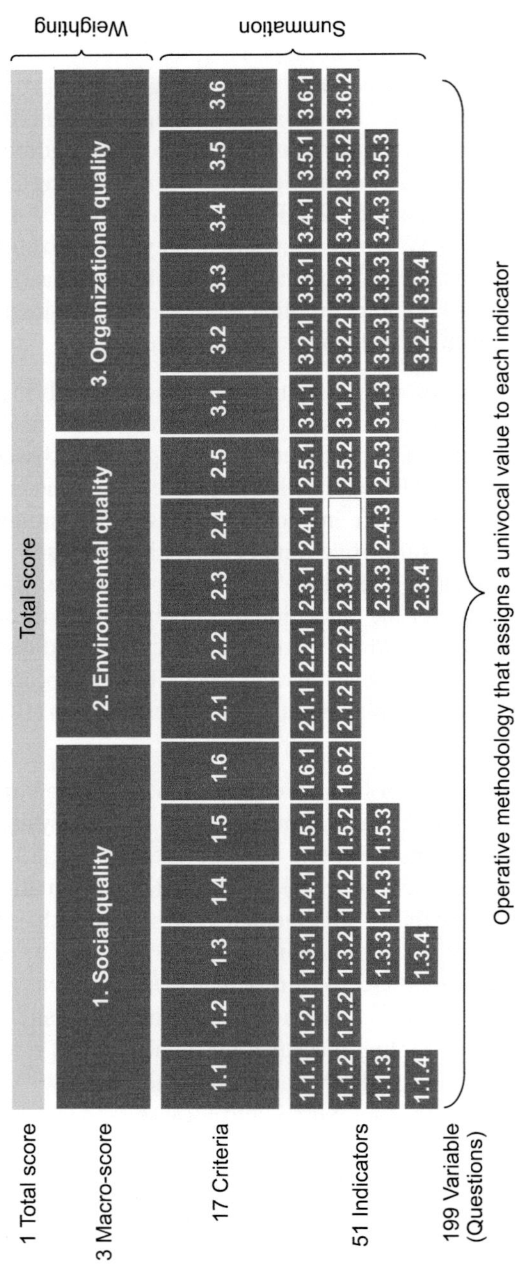

Fig. 3 Operative methodology.

Source: Authors

results provide a reference for targeted improvement of the hospitals (Brambilla et al., 2022). In addition, the national application of SusHealth V2 from Italy in Germany has also made attempts and provided improvements for adaptive applications in other regions.

The evaluation results, analyzed both within specific macro areas and for the entire building, afford a detailed or holistic perspective, illustrating the tool's capacity to provide nuanced insights into healthcare facility design and operation. This discussion underscores the tool's potential to significantly influence healthcare environments, enhancing user experience and accessibility through strategic design and operational improvements.

3.2 The Structure and Content of Design for All A.U.D.I.T.

Tool Framework: DfA A.U.D.I.T. evaluation framework introduces a tool designed to assess accessibility and inclusion within healthcare environments. The framework is structured around three principal categories (Physical, Sensory and Cognitive, and Social qualities), encompasses eight criteria, delineates 20 indicators, and specifies over 500 requirements to analyze the accessibility and inclusion of buildings shown in Figure 4 (Mosca et al., 2022). It facilitates the analysis of the impact of DfA strategy across various hospital areas, employing a rating system predicated on performance metrics. The tool is methodically segmented to represent different building environments, transitioning from external to internal spaces, including outdoor space, entrance, horizontal circulation, vertical circulation, waiting areas, refreshment areas, bathrooms, offices, inpatient department, outpatient, with particular attention to wayfinding across distinct building zones. This nuanced approach acknowledges the unique orientation needs at entrances versus corridors.

The first quality, Physical, integrates three criteria: usability, functionality, and safety/security. Usability ensures that all users can access and use the environment and products in a comfortable, easy, and equitable way, accommodating a wide range of abilities, such as incorporating rest areas within corridors to aid navigation over extended distances or materials of the surfaces to support usability. Functionality is exemplified by both the flexibility of furniture to adapt to individual needs and to functionality of space distribution, while the safety/security criteria focus on risk management and the perception of safety.

Sensory and Cognitive quality, the second principal quality, relates to wayfinding, understanding, and environmental standards. Wayfinding criteria advocate for diverse orientation methods, using visual cues and the building's layout as well as tactile and audio solutions for different sensory needs. The understanding criteria aim for intuitive communication of information regardless of the cognitive and sensory abilities of the users, with the support of aids such as colors, symbols, or digital devices. The environmental quality considers light, acoustics, air quality, and thermal comfort, ensuring a conducive atmosphere for all users.

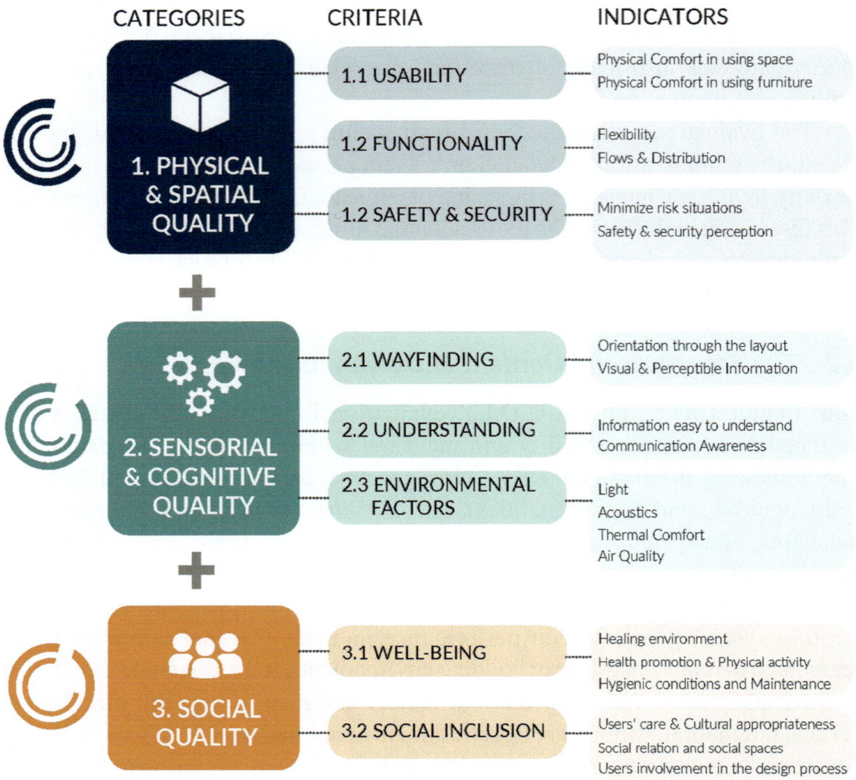

Fig. 4 Framework of A.U.D.I.T.
Source: Authors.

Finally, Social quality underscores the criteria of wellbeing and social inclusion. The first one includes a healing environment, physical activity promotion, hygienic conditions, and maintenance fostering well-being through the healing quality of the space. The latter includes indicators to fostering user relationships, cultural appropriateness, and emphasizing the significance of engaging various stakeholders in the design process. This comprehensive framework thus provides a structured approach to evaluating and enhancing healthcare environments, ensuring they are inclusive, functional, and conducive to the well-being of all users.

Scoring: The scoring system assesses requirements using a binary scale where the value of each of the indicators derives from the presence or absence of various requirements (Yes, No, N/A). The DfA quality assessment is the result of the fulfillment of the requirements defined for each building environment. Scores ranging from 1–3 reflect the requirements impact on achieving the specified indicator. The calculation of each indicator, and by extension the criteria to which various indicators are linked, employs a specific formula (Mosca and Capolongo,

2023). The aggregate score for each category is then determined by a weighted sum of the scores attained for each criterion. Consequently, the tool serves as a comprehensive assessment system and decision support system for designers and healthcare managers to improve the quality of hospitals for a wider range of users. It facilitates the evaluation of accessibility and inclusion in both projects and existing buildings, prioritizing interventions that enhance user well-being.

Tool Applications: The study details the tool's application in two private hospitals in Buffalo (U.S.A.) and Milan (Italy). Horizontal circulation scored highest in both hospitals (76% in Italy, 88% U.S.), while vertical circulation had the lowest score (46%) in the Italian hospital. The U.S. hospital had the lowest score in outdoor spaces. Design interventions are suggested for these areas. In UD criteria, the Italian hospital scored highest in environmental factors (84%) due to sustainable design, and lowest in social inclusion (54%). The U.S. hospital excelled in social inclusion (87%), but scored low in wayfinding (39%). Both hospitals had lower scores in wayfinding (59% Italy, 39% U.S.) due to layout and signs. The tool was able to provide specific feedback on UD outcomes as usability, orientation, and inclusion for different building environments. Therefore, the assessment with DfA AUDIT can be used to compare case studies or offer suggestions for building improvement.

To effectively communicate the results to decision-makers and designers, the study proposes various methods, including quantitative scoring for the entire building and qualitative reports that provide improvement suggestions, complemented by graphical evaluations for spatial analysis, available for both the overall building and focused areas. The tool is now available via a digital platform named Wideclusive ® (https://www.wideclusive.com/en/home), developed by the research group Design and Health Lab of Politecnico di Milano (Figure 5).

4. Discussion

4.1 Implications of Evaluating Sustainability with Specific Assessment Tools

SustHealth V2, characterized by its evidence-informed approach, utilizes a binary system and a weighting procedure grounded in a multi-criteria methodology. These features collectively form a robust decision support system that facilitates ongoing monitoring and control processes, ultimately aiming to enhance the quality of healthcare environments.

The tool's implementation strategies are designed to generate comprehensive reports, serving as the primary output. These reports provide an in-depth assessment of the facility's specific quality areas, offering healthcare organizations actionable insights to formulate improvement, checking, and planning strategies. This iterative process underscores the tool's capacity to drive significant quality enhancements within healthcare settings (Buffoli et al., 2015; Devlin and Arneill, 2003).

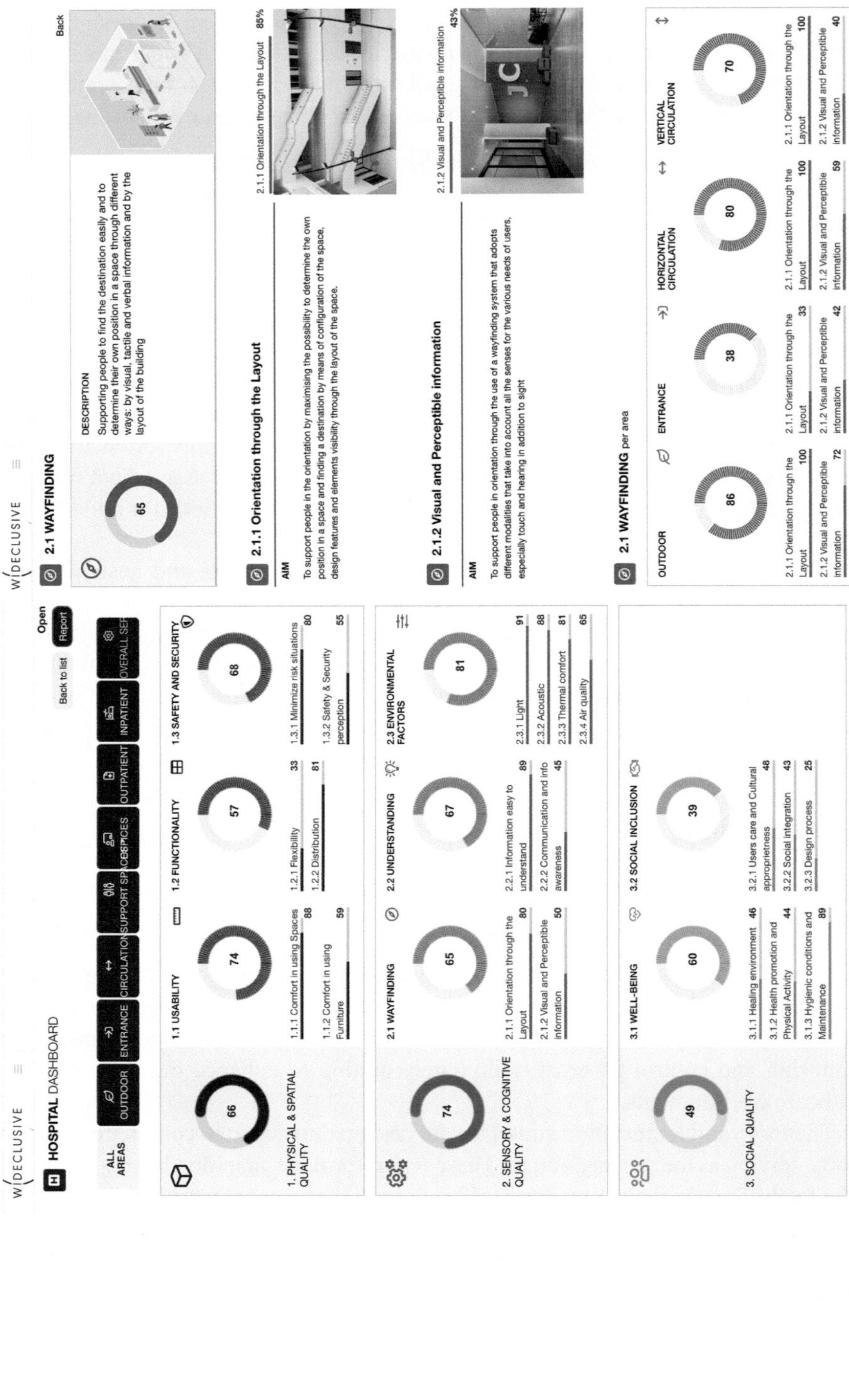

Fig. 5 Website interface of Wideclusive ®
Source: Authors.

The core of the tool's innovation and impact is its foundation in evidence-based practices, employing a simplified binary system to streamline the evaluation process. This approach, coupled with a multi-criterion weighting procedure, equips healthcare administrators with a detailed decision support system. Such a system not only aids in the effective monitoring and control of quality improvement initiatives but also contributes to the overarching goal of fostering better hospitals for the future (Brambilla and Capolongo, 2019) .

Furthermore, the development of the tool involves transforming it into an application that enhances its functionality with automated features for more efficient data collection. This advancement has been piloted in 15 European hospitals, highlighting the need for automated data collection mechanisms to optimize the tool's efficacy. The result of this development is not just a detailed report, but a set of improvement strategies tailored to each hospital's unique needs and circumstances (Brambilla et al., 2022).

This progression towards automation and app development signifies a leap forward in how healthcare environments can utilize technology to assess and improve quality systematically. By using automatic data collection and analysis, the tool is expected to provide healthcare organizations with more precise and actionable feedback, paving the way for targeted and effective quality improvement initiatives.

The SustHealth V2 tool represents a significant contribution to the healthcare quality improvement landscape. Its evidence-informed framework, combined with innovative binary and multi-criteria decision support capabilities, offers a comprehensive approach to enhancing hospital quality. As the tool evolves, its increased automation and application-based interface are expected to further streamline the quality improvement process, offering healthcare facilities with advanced means to achieve and maintain high standards of care.

4.2 Implications of Evaluating Inclusion with Specific Assessment Tools

DfA A.U.D.I.T. tool, developed by a doctoral researcher from the Design and Health Lab of Politecnico di Milano (Mosca and Capolongo, 2023), represents a significant advancement in the field of inclusive design. This tool, underpinned by the principles of DfA, offers a structured framework for evaluating and improving healthcare environments to cater to the diverse needs of all users. By incorporating a multi-criteria evaluation framework that assesses physical, sensory, cognitive, and social qualities through a comprehensive set of criteria and indicators, DfA A.U.D.I.T. facilitates a holistic approach to designing healthcare facilities (Mosca et al., 2022). Moreover, the application of DfA A.U.D.I.T. in pilot studies, such as those conducted in private hospitals, validates its effectiveness and applicability. The objective data collected through the tool's indicators and criteria, complemented by subjective user experience surveys, offers a nuanced understanding of the healthcare environment's impact on users.

The research group Design and Health Lab has developed a digital platform for the tool called Wideclusive ®, showcasing the technological innovation behind this project. It enables rapid data collection and benchmark creation, allowing for the efficient analysis of healthcare facilities against established standards. This capability not only streamlines the evaluation process but also provides a robust database for benchmarking, which is critical for identifying priority areas for intervention and enhancing overall well-being and customer satisfaction. The flexibility of DfA A.U.D.I.T. tool's structure ensures its applicability across a variety of building types, although its initial focus has been on healthcare facilities.

The implications of DfA A.U.D.I.T. for healthcare facilities are profound. By providing a comprehensive evaluation framework, healthcare designers and managers are equipped with the insights needed to prioritize interventions that significantly enhance user experience, accessibility and inclusion in the design and renovation process. The tool's emphasis on a rating system based on performance, considering a wide array of criteria from usability to environmental quality, ensures that designs are not only functional but also inclusive and conducive to healing. DfA A.U.D.I.T., with its comprehensive evaluation framework and innovative technological platform, offers a promising avenue for enhancing the design and operation of healthcare facilities.

4.3 Limitations and Future Developments of the Tools

SustHealth V2 is used in the practical study both in Italy and Germany, and the outcome shown limitations and possible improvements. Due to the small size of hospital as samples, when analyzing different types of hospitals, such as clustering hospitals according to ownership and number of patient beds, the size of each segmented cohort is not enough to show a strong link (Brambilla et al., 2022). For enhanced rigor, future inquiries should extend the application with wider sample sizes across broader demographic cohorts and varied environmental conditions, encompassing diverse healthcare infrastructures. The impact of social, economic, and infrastructural variables, such as hospital size, on outcomes, necessitates a multifaceted approach in subsequent analyses.

A similar limitation is shown in the application of the DfA A.U.D.I.T. To realize the idea of using assessment tools to support evidence-based design, more cases should be evaluated and analyzed. DfA A.U.D.I.T.'s current platform, Wideclusive ®, has been designed to enable the collection of information more broadly and programmatically for a wider range of applications. Wideclusive ® utilizes a web app and a digital checklist processed by an evaluator, laying the groundwork for more sophisticated data collection methods. Future developments aim to incorporate design software such as Building Information Modeling (BIM) and hardware like sensors to automate the collection of quantitative indicators. This approach will facilitate the assessment of material characteristics and spatial parameters, akin to BIM model checking, tailored to evaluate DfA features.

Although the SustHealth V2 and DfA A.U.D.I.T. are developed specifically for improving the quality of healthcare facilities, they could inspire the architecture design during the project phase concerning the topic of sustainability and design for all. The method of development and content of these two tools also provide references for construction assessment tools in other fields. After the adoption of the aim building type, there is a possibility for SustHealth V2 and DfA A.U.D.I.T. to be used for the quality evaluation of other type buildings.

Conclusion

This evaluation of the healthcare environment reinforces the value of integrating evidence-based tools in the healthcare sector, not only to enhance the quality and sustainability of healthcare environments but also to ensure that strategic decisions are informed by comprehensive and rigorously analyzed data. For example, the comprehensive analysis of data derived from assessments in the SustHealth V2 tool is crucial for supporting evidence-based and measurement-driven decision-making. This analytical approach enables the application of the evaluation tool, as well as similar instruments, across broad samples, facilitating strategic decision-making processes that significantly contribute to policy formulation across various domains.

The experience derived from this research underscores the potential for such tools to influence policymaking and emphasizes the need for a collaborative and multidisciplinary approach to tackle complex challenges within healthcare settings effectively. The insights gained from tools have further motivated the development of systematic evaluation and dissemination strategies at both regional and national levels, aiming to address emerging challenges within healthcare environments, such as those presented by the COVID-19 pandemic. This endeavor highlights the importance of adopting a holistic perspective that acknowledges the necessity for cooperation and interdisciplinary collaboration across different sectors, with a particular emphasis on the healthcare industry.

From the aspect of improving all user satisfaction and developing user-center built environment, the evaluation tool of healthcare environment contributes to the social sustainability. The evaluation of DfA quality within healthcare environments, encompassing physical, sensorial, cognitive, and social dimensions, underscores the necessity to address diverse user needs. These qualities assessed in the tool serve as the principal objectives, highlighting the intrinsic link between user requirements and the broader perspective of the built environment, a connection that is particularly pronounced in healthcare settings.

The employed methodology demonstrates the efficacy of segmenting complex strategic issues into manageable components, as evidenced by the dissection of DfA into specific criteria and indicators. This approach enables the application of a multifaceted methodology, leveraging various methods to gather comprehensive information from disparate sources throughout the design process. The critical review of ideas, facilitated by engaging experts and users in discussions to explore diverse viewpoints is a scientific and rigorous methodology support for improving the quality of healthcare environment to reach the goal of human-centered and social sustainable design.

The development and implementation of a user-centered tool not only foster a reflective consideration of the end-users' needs but also guide the optimization of the design and services. By critically analyzing the assessment results, the application of tool for healthcare environment provides clear directions for improvement and design recommendations that can be referenced, ensuring its practicality and effectiveness for the intended beneficiaries. The assessment tool emphasizes the importance of a comprehensive, user-focused methodology in enhancing healthcare facility design and operation, promising significant advancements in user satisfaction and accessibility, and ultimately contributing to more inclusive healthcare environments. Further developments will include the application of the tool in a wide variety of healthcare case studies in order to enable sample consistency and statistical analysis.

In conclusion, this paper reinforces the necessity of integrating evidence-based tools like SustHealth V2 and A.U.D.I.T in healthcare settings. Such tools play a pivotal role in enhancing the quality, sustainability, and user-center experience of healthcare environments, guiding strategic decisions based on multiple data sources, and promoting a collaborative, multidisciplinary approach to overcome the sector's challenges.

References

Astley, P., Capolongo, S., Gola, M. and Tartaglia, A. (2015). Operative and design adaptability in healthcare facilities. *TECHNE-Journal of Technology for Architecture and Environment*, 162–70.

Brambilla, A., Apel, J.M., Schmidt-Ross, I., Buffoli, M. and Capolongo, S. (2022). Testing of a Multiple Criteria Assessment Tool for Healthcare Facilities Quality and Sustainability: The Case of German Hospitals. *Sustainability*, *14*(24), 16742. https://doi.org/10.3390/su142416742.

Brambilla, A. and Capolongo, S. (2019). Healthy and Sustainable Hospital Evaluation: A Review of POE Tools for Hospital Assessment in an Evidence-Based Design Framework. *Buildings*, *9*(4), Article 4. https://doi.org/10.3390/buildings9040076.

Brambilla, A., Lindahl, G., Dell'Ovo, M. and Capolongo, S. (2020). Validation of a multiple criteria tool for healthcare facilities quality evaluation. *Facilities*, *39*(5/6), 434–47. https://doi.org/10.1108/F-06-2020-0070.

Brambilla, A., Lindahl, G., Dell'Ovo, M. and Capolongo, S. (2021). Validation of a multiple criteria tool for healthcare facilities quality evaluation. *Facilities*, *39*(5/6), 434–47. https://doi.org/10.1108/F-06-2020-0070.

Buffoli, M., Capolongo, S., di Noia, M., Gherardi, G. and Gola, M. (2015). Healthcare sustainability evaluation systems. *Improving Sustainability during Hospital Design and Operation: A Multidisciplinary Evaluation Tool*, 23–29.

Cambra-Rufino, L., Brambilla, A., Paniagua-Caparrós, J.L. and Capolongo, S. (2021). Hospital Architecture in Spain and Italy: Gaps Between Education and Practice. *HERD: Health Environments Research & Design Journal*, *14*(3), 169–81. https://doi.org/10.1177/1937586721991520.

Capolongo, S., Gola, M., Noia, M.D., Nickolova, M., Nachiero, D., Rebecchi, A., Settimo, G., Vittori, G. and Buffoli, M. (2016). Social sustainability in healthcare facilities: A rating tool for analysing and improving social aspects in environments of care. *Annali Dell'Istituto Superiore Di Sanità*, *52*(1), Article 1.

Corrente, S., Figueira, J.R. and Greco, S. (2019). A new scaling MCDA procedure putting together pairwise comparison tables and the deck of cards method. *arXiv Preprint arXiv:1904.01315.*

Department of Health and Social Care. (2020, October 2). *PM Confirms £3.7 Billion for 40 Hospitals in Biggest Hospital Building Programme in a Generation.* GOV.UK. https://www.gov.uk/government/news/pm-confirms-37-billion-for-40-hospitals-in-biggest-hospital-building-programme-in-a-generation.

Devlin, A.S. and Arneill, A.B. (2003). Health care environments and patient outcomes: A review of the literature. *Environment and Behavior, 35*(5), 665–94.

Dolcini, M., Brambilla, A., Gola, M. and Capolongo, S. (2023). Health and well-being key performance indicators in corporate sustainability disclosure. A review of sustainability reports from a sample of major European companies. *Acta Biomedica Atenei Parmensis, 94*(S3), e2023132. https://doi.org/10.23750/abm.v94iS3.14334.

European Institute for Design and Disability. (2004). *The EIDD Stockholm Declaration.* https://dfaeurope.eu/wordpress/wp-content/uploads/2014/05/stockholm-declaration_english.pdf.

Figueira, J. and Roy, B. (2002). Determining the weights of criteria in the ELECTRE type methods with a revised Simos' procedure. *European Journal of Operational Research, 139*(2), 317–26.

Figueiro, M. (2008). Sustainable lighting for healthcare facilities: More than just lumens per watt. *Journal of Green Building.* https://doi.org/10.3992/jgb.3.1.74.

Haapio, A. (2012). Towards sustainable urban communities. *Environmental Impact Assessment Review, 32*(1), 165–69. Scopus. https://doi.org/10.1016/j.eiar.2011.08.002.

Jaušovec, M. and Gabrovec, B. (2023). Architectural Evaluation of Healthcare Facilities: A Comprehensive Review and Implications for Building Design. *Buildings, 13*(12), 2926. https://doi.org/10.3390/buildings13122926.

Lega, F., Mauri, C. and Prenestini, A. (2011). L'ospedale tra presente e futuro: Analisi, diagnosi e linee di cambiamento per il sistema ospedaliero italiano. *EGEA spa.*

Mosca, E.I. and Capolongo, S. (2018). Towards a Universal Design Evaluation for Assessing the Performance of the Built Environment. *In: Transforming our World Through Design, Diversity, and Education*, 771–79. IOS Press. https://doi.org/10.3233/978-1-61499-923-2-771.

Mosca, E.I. and Capolongo, S. (2023). Design for All AUDIT (Assessment Universal Design & Inclusion Tool). A tool to evaluate physical, sensory-cognitive and social quality in healthcare facilities. *Acta Bio-Medica: Atenei Parmensis, 94*(S3), e2023124–e2023124.

Mosca, E.I., Herssens, J., Rebecchi, A., Strickfaden, M. and Capolongo, S. (2019). *Evaluating a Proposed Design for All (DfA) Manual for Architecture*, 54–64.

Mosca, E., White, J., Steinfeld, E. and Capolongo, S. (2022). Designing Hospitals Through the Lens of Universal Design. An Evaluation Tool to Enhance Inclusive Healthcare Facilities. *In: Studies in Health Technology and Informatics, 297.* https://doi.org/10.3233/SHTI220857.

PNRR, gli obiettivi e la struttura. (2021, April 23). www.governo.it. https://www.governo.it/it/approfondimento/pnrr-gli-obiettivi-e-la-struttura/16702.

Ricciardi, W. (2015). La tempesta perfetta: Il possibile naufragio del Servizio sanitario nazionale: Come evitarlo? *Vita e pensiero.*

Sdino, L., Brambilla, A., Dell'Ovo, M., Sdino, B. and Capolongo, S. (2021). Hospital Construction Cost Affecting Their Lifecycle: An Italian Overview. *Healthcare, 9*(7), Article 7. https://doi.org/10.3390/healthcare9070888.

Ulrich, R.S., Berry, L.L., Quan, X. and Parish, J.T. (2010). A conceptual framework for the domain of evidence-based design. *HERD: Health Environments Research & Design Journal, 4*(1), 95–114.

Ulrich, R.S., Zimring, C., Zhu, X., DuBose, J., Seo, H.-B., Choi, Y.-S., Quan, X. and Joseph, A. (2008). A Review of the Research Literature on Evidence-Based Healthcare Design. *HERD: Health Environments Research & Design Journal, 1*(3), 61–125. https://doi.org/10.1177/193758670800100306.

White, J. and Mosca, E.I. (2022). Developing Innovative Solutions for Universal Design in Healthcare and Other Sectors. *Stud. Health Technol. Inform., 297*, 340–47.

10 The Convergence of Resilience Culture and Digital Transformation in the Cities of Future

The Case Study of Milan and Rome

Stefania Stellacci[1*] and Giovanni Borsoi[2]

1. Introduction

Over 75% of the world population is projected to reside in urban areas by 2050. Therefore, contemporary cities and their lively communities play a pivotal role to either generate and mitigate natural disasters and human-made risk impacts along different phases of emergency management, such as mitigation, preparedness, response, and recovery (Wisner and Adams, 2002; UN 2015, 2019).

Within this challenging and fast-changing context, urban sustainability can be achieved through integrated future-oriented policy actions that impact liveability, circularity, and resilience (SNPA, 2022). To assess the level of urban resilience, multidimensional frameworks based on interdepended indicators have been developed in the last few years (UNISDR, 2012; Desouza and Flanery, 2013; Manca et al., 2017; Saja et al., 2019; SNPA 2022). One of the most comprehensive and well-structured framework related to natural, social, and built environments has been developed by The Rockfeller Foundation and Arup (2016), where the urban resilience depends on four key dimensions: i) health and well-being; ii) economy and society; iii) infrastructure and environment; and iv) leadership and strategy. Each dimension is aligned with three specific objectives. This framework comprises 52 resilience indicators, evaluated using 156 questions that draw on both qualitative and quantitative data. The answers to these questions are organized based on the 12 objectives, the importance of which depends on the urban context and the specific challenges of the urban environment. A city resilient to climate change should embody qualities of robustness, redundancy, flexibility, resourcefulness, reflectiveness, inclusivity, and integration, as defined in the *City Resilience Index* by The Rockefeller Foundation and Arup (2016).

1 ISTAR-IUL, Information Sciences, Technologies and Architecture Research Center, Iscte, University Institute of Lisbon

2 CERIS, Civil Engineering Research and Innovation for Sustainability, Instituto Superior Técnico, University of Lisbon
Email : giovanni.borsoi@tecnico.ulisboa.pt

* Corresponding author: Stefania_Stellacci@iscte-iul.pt

The *Resilience Maturity Model*, conceived within the EU-financed *Smart Mature Resilience* (SMR) project,[1] defines strategies to enhance cities resilience considering four main groups (leadership and governance; infrastructure and resources; preparedness; and cooperation). Ten policies' sub-dimensions and the respective effort indicators and result indicators are identified under these four categories. This model defines a pathway through five stages to achieve the "resilience maturity" (starting, moderate, advanced, robust, and verTebrate[2]). This model thus supports the decision-making process of different stakeholders for establishing appropriate policies, facilitating the progression to the next stage of maturity.

In parallel to studies focused on resilience frameworks, academics, and practitioners (e.g., Kudo, 2018; Vermiglio et al., 2020) have also explored in recent years how the progress and pervasive adoption of digital tools and smart technologies can enhance urban resilience to the escalating frequency and severity of disaster risks (IPCC, 2023). The debate involving urban dwellers, city planners, environmentalists, ecologists, city policymakers, investors, and other key players in urban management and IT-development remains open and requires further exploration to better inform future urban and environmental decision-making processes. On the one hand, it is crucial to promptly address disruptive events, prevent and reduce their compounding and cascading effects, and adapt both environment features and human behaviours. On the other hand, analysis and design framework should be overall refined by incorporating approaches and tools that have so far been little experimented with and proven in real cases (e.g., *vide* criticisms identified by Bugs et al., 2010).

In this study, we provide a summary of the comparative assessment of the resilience approach, examining the report published by *Sistema Nazionale per la Protezione dell'Ambiente* (SNPA, 2022) crossed by reflections arising from the analysis of 100 Resilience City (100RC) initiative. 100RC was launched by The Rockefeller Foundation in 2013 (Clark et al., 2018; Comune di Roma, 2018; Galderisi et al., 2020). With a focus on two italian cities that are members of 100RC (Milan and Rome), the wide discourse around resilience culture and digital transformation is connected to a multiple-indicators-based evaluation of urban resilience levels. In examining these two cities, we aim to contribute to the broader discourse on the future of European cities.

Reports and other documents on these two urban contexts, which can be unknown by non-English academics, was analysed in this work and related to current and future strategic actions around enriched resilience culture. We also focus on how technological advancement in digital architectural applications can positively impact the social, economic, environmental, and institutional dimensions

---

1 https://smr-project.eu/home/

2 ***starting***: starting with local (departmental) resilience plans; ***moderate***: integration of local (departmental) resilience plans; ***advanced***: implementation of the integrated (holistic) resilience plan; ***robust***: internationalizing resilience; ***verTebrate***: leading resilient city.

of urban areas, while possible risks and limitations on these applications are identified.

This chapter is structured into four sections. Firstly, we delve into the complex domain at the intersection of urban planning and disaster risk management, examining both the polysemic concept of urban resilience and a resilience framework. Secondly, we provide a brief summary of the comparative analysis conducted by the Italian information infrastructures in national territory to evaluate the resilience level to climate change of two selected Italian cities (SNPA, 2022). We focus on Milan and Rome, two high-density cities prone to multiple natural and human-induced threats. In the third section, we examine data from the 100RC project, which included Milan and Rome as selected Italian members. Finally, we reflect on the future trajectories of the European urban agenda aimed at addressing some criticalities investigated in this study and narrowing both the digital divide and digital exclusion.

2. Urban Resilience in Literature: A Brief Overview

Resilience is a polysemic and slippery concept that is used in different domains, from ecology (Holling, 1973), psychology, or social sciences, physics, and engineering, to the EU-science policy interface (Baravikova et al., 2020). Comfort et al. (2001), Park (2011), and Reghezza-Zitt et al. (2012) express scepticism about the excessive use of this buzzword. While several authors attribute the introduction of this concept to Holling (1973), its use can be traced back several centuries earlier to the Latin word *resilio* that means "jump back" or "bounce back" (Manyena, 2006:, 433; Alexander, 2013: 2708). In the field of biology and ecology, Holling draws attention to this concept, intended as the ability of an ecological system to absorb changes or disturbances (that can be shocks and slow burn processes), without significant alterations and "still maintain the same relationships between populations or state variables" (Holling, 1973: 14).

Alexander (2013) comes across an etymological journey from AD 35 to the 2010s of the term 'resilience', between dichotomous terms, stability dynamism, or dynamic equilibrium *vs.* (homeostasis) and evolution. "As a concept, resilience shows promise in that it encourages the researcher to bridge the 'shear zone' between (dynamic) adaptation and (static) resistance." (Alexander, 2013: 2714). Figure 1 shows the migration of the term resilience from different fields and the two main approaches: 'engineering' or "bouncing back" approach *vs.* 'evolutionary' or "bouncing forward" approach. Engineering resilience refers to the ability of systems to return to a stable steady-state, i.e., its equilibrium state, after a perturbation. Resilience implies an evolutionary character, a non-stable equilibrium, as conceived by Davoudi (2012). A shift from the engineering to evolutionary approach has emerged, as stated by Baravikova et al. (2020). However, some criticism may arise due to the omission of the political dimension.

Furthermore, 'resilience' has been employed in environmental and social sciences, and in disaster risk reduction (DRR), with a vast range of interpretations.[3] In cultural ecology, this concept is the ability to adapt to environmental extremes. When considering a socio-ecological and urban approach, resilience is defined as the amount of change the system can undergo and still retain the same controls on function and structure; the degree to which the system is capable of self-organization; and the ability to build and increase the capacity for learning and adaptation (Holling and Walker, 2003). In recent years, human adaptive behaviour has been widely studied, since a resilient society embodies both absorptive, adaptive, and transformative capacities (Manca et al., 2017).

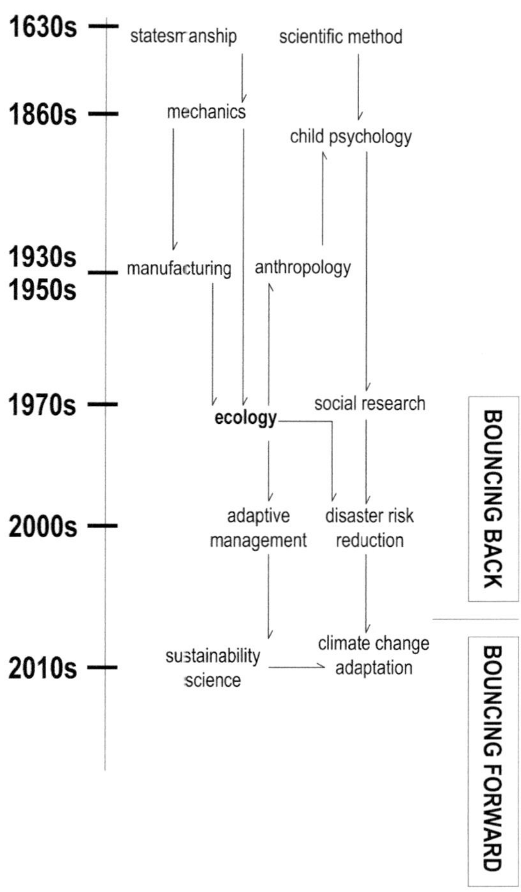

Fig. 1 The concept of resilience over time, adapted from Alexander (2013: 2714).

[3] see e.g., Desouza and Flanery (2013); Meerow, Newell & Stults (2016); Figueiredo et al. (2018).

Academics have emphasized the necessity of establishing stronger connections between urban resilience theory and practical applications (Coaffee et al., 2018). In addition to the need of a clearer conceptualization of "urban resilience" and operationalization, political tensions arise on how to reconcile wider applicability and local specificity, as stated by Baravikova et al. (2020), due to the lack of operationalization efforts and absence of normativity of urban resilience thinking. Thus, the insufficient integration of resilience thinking with the political aspects of risk leads to a significant limitation of urban resilience in the current practice. Bahadur and Tanner (2014:202) also emphasize "a failure to address trade-offs, and its epistemological bias towards technical responses". This means an inclination to frame theoretically, and tackle emerging challenges using technical approaches and knowledge, while the social dimension is overlooked. Conversely, the 100 Resilient Cities (100RC) project—examined with more detail in the next sections—focuses on vulnerable social sectors and embraces a more evolutionary approach: "[c]ity resilience describes the capacity of cities to function, so that the people living and working in cities—particularly the poor and vulnerable—survive and thrive no matter what stresses or shocks they encounter" (The Rockefeller Foundation and Arup 2015: 11).

Resuming the concept of resilience in an evolutionary sense, it seems advantageous to implement adaptive management, which involves the selection of a strategy that can be continually adjusted to enhance performance as understanding improves and information about the future expands.

3. Resilience Framework and Multisector Indicators for Urban Resilience to Climate Change

Under the broad umbrella of 'resilience', urban resilience to climate change is a relevant benchmark within current environmental policies. By analysing an extensive body of literature e.g., UNISDR (2008), Cutter, Ash and Emrich (2014), The Rockefeller Foundation and Arup (2016), and UN-Habitat (2016a, 2016b, 2017), Figueiredo et al. (2018) provide a comprehensive resilience framework. A wide spectrum of indicators and sub-drivers is identified for measuring the resilience levels in cities across social, economic, environment, and institutional sectors. The social dimension focuses on creating an inclusive and cohesive society, promoting citizen networks, and providing fair and broad access to information, opportunities, services, and amenities. The economic dimension involves promoting industry diversification and innovation. Additionally, the environmental sphere assesses the sustainability of urban development, the availability of reliable infrastructure, and the adequacy of natural resources. Finally, the institutional sector necessitates clear leadership, a future-oriented vision, a set of long-term funding resources, synergies and collaboration across different government levels, as well as an open and participatory governance approach. It is important to note that these indicators are not intended to be universal or unalterable over time, in line with transformational approaches to resilience. They may not be applicable to all real-world scenarios, even though similar risks impact diverse urban environments in distant geographic areas.

For each sector and its respective indicator, seven qualities (or capacities) of resilient cities, as outlined in the above-mentioned *City Resilience Index Reference*, are associated as follows: (i) robust; (ii) redundant; (iii) flexible; (iv) resourceful; (v) reflective; (vi) inclusive; and (vii) integrated. The first two qualities (i, ii) fall within the prevention preparedness and pre-existing phases (see Figure 2, bottom). The following three qualities (iii, iv, v), based on outcome-oriented approach, are relevant to response and recovery. The last two qualities (vi, vii) focus on building together and are process-oriented. In the report published by *Sistema Nazionale per la Protezione dell'Ambiente* (SNPA, 2022), these attributes of the cities are indicated as fundamental for preventing any trauma and stress for people and natural/built environments, and they can be monitored in connection with diverse assets, systems, behaviours, and practices implemented within the city (SNPA, 2022).

To define city-specific indicators, we should bear in mind that the social sphere covers all times and phase, emphases, except for (v) and (vii) (Figure 2, bottom). Conversely, we should consider only two main set of indicators (those related to social and environmental sectors) to assess if the city is redundant, and three institutional-related indicators, environmental and institutional-related parameters in the case of a 'reflective' city. Similarly, the integrated city will be based on environmental and institutional spheres. The qualities of resilient cities (robustness, resourcefulness, and inclusivity) depend on the interplay of indicators related to three spheres, i.e., the robustness involves the social, environmental, and institutional spheres; the resourcefulness involves social, economic, and institutional; inclusivity regards all-spheres related indicators.

Figure 2 shows the conceptual framework for urban resilience in a radial diagram, elaborated based on the data published by Figueiredo et al. (2018: 59–66), which shows 68 indicators across the four abovementioned sectors. These indicators range from quantitative outcomes or inputs, quantitative outputs, to quantitative or qualitative process. The qualitative process involves the institutional sector, while quantitative outcomes can be found in the economic-based indicators, and both quantitative outcomes and outputs in the social-based indicators. In a nutshell, the broad framework of the urban resilience relies on the interdependence of those indicators that evaluate ecosystemic and human functions.

Although being based on the holistic and whole-of-society approach, this resilience framework features some limitations. These arise from the inherent difficulties to capture all dimensions of the urban resilience using 'measurable' indicators. As an example, no indications on how to measure the quality of the job are referred in this framework. Furthermore, it is worth mentioning the influence of policy is hardly disentangled from the influence of confounding factors, thus the "effectiveness of a given policy cannot be determined by indicators alone" (Figueiredo et al., 2018: 44).

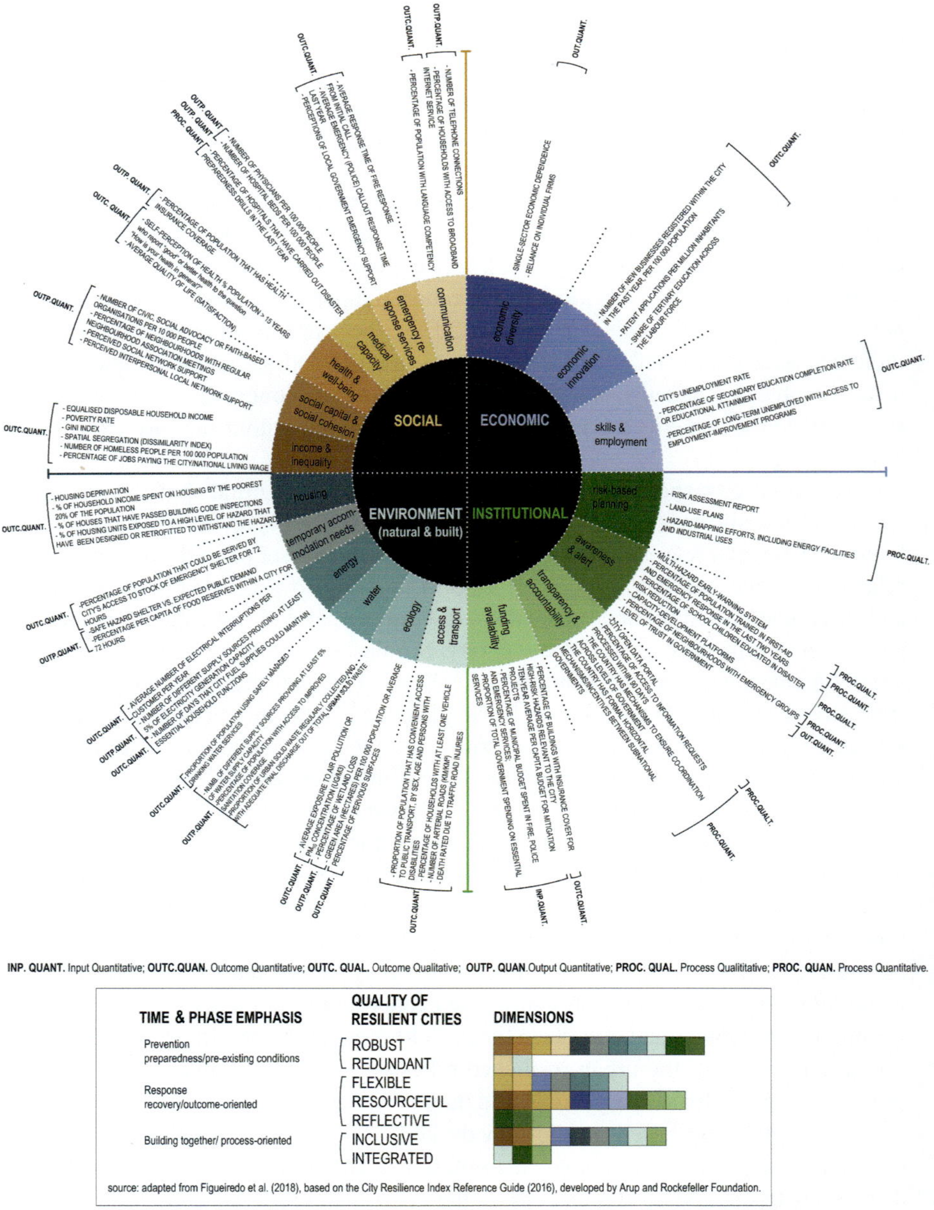

Fig. 2 Indicators to measure urban resilience, using information from Figueiredo et al. (2018:59–66) (Authors' own elaboration).

As indicated by Peters et al. (2016) and restated by Figueiredo et al. (2018), it would be beneficial to establish consistency across the global framework for increasing resilience of the cities. Such consistency could enhance the effectiveness of sustainable strategy development since coordinated and unified strategies tend to be more successful than those implemented in isolation. As a matter of fact, it is proven that the enhancement of resilience without considering the overall framework may adversely affect resilience itself at other levels. As stated by Bahadur and Tanner (2014), this could occur, for instance, by diverting resources away from other systems or by exploiting different groups of people. Instead of dedicating time and other resources to achieve the goals and targets individually, national, regional and city authorities should focus on defining an integrated implementation plan across different sectors, with resilience as the overarching objective that aligns all intentions.

Other guiding principles are defined in the *Sendai Framework for Disaster Risk Reduction* 2015–2030,[4] e.g., the engagement from all sectors of the society, the central role of national governments in DRR, and the need of collaboration and support from developed countries and partners to developing countries.

3.1 Indicators of the Urban Resilience and Respective Targets in Italian Cities

Focusing on the Italian context, 56% of the population resides in urban centres (2021) and the urbanized areas are increasing. Italy has approved its own *National Adaptation Strategy* (NAS) in 2021, yet with a low operationalization. Due to its decentralized governance system, significant responsibilities rest with regional governments and local authorities. Different criticalities and risks, considering factors such as fragility, vulnerability, and exposure, can be observed across the national territory, both in urban and regional contexts.[5] This is also related to the distinctive peculiarities and complexity of the natural and heritage landscape in Italy, including numerous sites listed in the UNESCO World Heritage list. Additionally, it also depends on the diverse economic and infrastructural features in each city and region. Other intangible assets, including patterns of human behaviour among local stakeholders, as well as diverse technical, administrative, and financial capacities dedicated to establishing pathways for climate actions, vary across the territory and also play a significant role in risk management.

The urban sustainability can be achieved through strategic actions that impact liveability, circularity, and resilience to climate change, as identified in the report recently provided by *Sistema Nazionale per la Protezione dell'Ambiente* (SNPA, 2022).[6] A series of targets linked to liveability, circularity, and resilience

---

4 https://www.undrr.org/monitoring-sendai-framework.

5 See the report on six cities in Italy, including Milan and Rome. https://www.cmcc.it/it/rischio-clima-citta-2021.

6 *National System for Environmental Protection* (https://www.snpambiente.it/) was established in 2017 to create a network of all institutions involved in environmental issues to ensure consistency and effectiveness in public control actions. It oversees the environmental condition, regulates sources and factors of pollution, conducts environmental research, offers technical and scientific assistance to state, regional, and local entities, and manages the collection, organization, and dissemination of environmental data.

to be accomplished by 2030 are set at the national level. Among those, a set of resilience targets regard the reduction of mortality and improvement of people health (*Target 3.4*), water efficiency improvement across all sectors (*Target 6.4*), and the improvement of sustainable and inclusive urban planning (*Target 11.3*).[7] To achieve *Target 3.4*, it is necessary to increase green areas in the municipal area to mitigate heatwaves, especially during the summer (May 15–September 15), and reduce the risk of floods and inundations. To achieve *Target 6.4*, specific actions to solve water scarcity and water waste need to be timely implemented to enhance local supply chains. To achieve *Target 11.3*, the reduction of concrete-based roofs, bituminous pavements or roads, or other impermeable surface can reduce soil consumption, whereas vegetated and permeable surfaces can reduce both the risk of floods and mitigate urban heat islands intensity.

Shifting the focus from the general pathway defined by Figueiredo et al. (2018) to a context-based framework applied to a group of cities in Italy,[8] five dimensions regarding urban resilience to climate change need to be examined: (i) land and territory; (ii) water; (iii) health; (iv) green infrastructure; and (v) socio-demographic structure. Within each of these dimensions, a set of interrelated drivers and indicators are analysed in the report by SNPA (2022) to assess the level of resilience in selected Italian cities. Ultimately, this set of indicators identifies key aspects crucial for successful strategy emergency management, based on a local city-centric approach. Geographic, territorial, and environmental data are processed by the *Istituto Superiore per la Protezione e la Ricerca Ambientale* (ISPRA) and SINA. The data range from cartographic or thematic maps to census data, and the time threshold of this dataset varies from 2012–2020 (as detailed in Figure 3).

Before delving into the analysis of trends in two samples (Milan and Rome), a few preliminary observations are necessary. Firstly, to have a clear broad picture, the three targets related to urban resilience to climate change (mortality and human health, water management, and urban planning) in this context cannot be considered independently from those related to other key indicators as liveability and circularity.[9] Secondly, regarding urban resilience indicator, one may observe

---

[7] "Target 3.4: By 2030, reduce premature mortality from non-communicable diseases by one-third through prevention and treatment and promote mental health and well-being. Target 6.4: By 2030, substantially increase water efficiency in all sectors and ensure sustainable withdrawals and supply of freshwater to address water scarcity, significantly reducing the number of people suffering from water scarcity. Target 11.3: By 2030, increase inclusive and sustainable urbanization, as well as the capacity for participatory and integrated planning and management of human settlements in all countries" (own translation).

[8] Twenty municipalities in Italy were chosen based on: (i) nationwide coverage and uniform distribution; (ii) demographic clusters (>500,000, >100,000, and <100,000 inhabitants); (iii) diversity of territorial and climatic features (coastal, plain, hilly, mountainous urban areas), i.e., Torino, Aosta, Genova, Milano, Bolzano, Trento, Venezia, Trieste, Bologna, Firenze, Perugia, Ancona, Roma, L'Aquila, Campobasso, Napoli, Bari, Potenza, Catanzaro, Palermo, and Cagliari.

[9] "The assignment of indicators to a specific interpretative key was decided based on their prevailing meaning, but it should be noted that they often have a cross-cutting relevance that goes beyond the classification made." (SNPA, 2022: 17) (own translation).

the relevance given to two specific contextual indicators, i.e., water (and its overuse) and urban heat island (and its impact on both the physical environment and people well-being).

As an example, a low resilient city will have a high area of impermeable surfaces to absorb significant quantities of water, being enable to mitigate urban heat islands intensity as well as prevent the risk of floods and inundations. In fact, the intensification and prolonged duration of heavy precipitation events, coupled with hydrogeological instability in the region and worsened by the consequences of uncontrolled soil consumption, pose an escalating threat to the population living in urban areas in Italy. Furthermore, the impact on health in terms of summer-related mortality is also high in low resilient urban environments. This reveals a hazard-specific framework tailored to the specific threats identified in Italian cities, whose relevance compared to all-hazard framework is stressed by Saya et al. (2019).

3.1.1 Resilience to climate change in Milan and Rome

Figure 3 summarizes the fluctuations of the urban resilience indicators in Milan (1.4 million inhabitants;[10] area: 182 Km^2) and Rome (2.8 million inhabitants;[11] area: 1,287 km^2), based on the report published by SNPA (2022). The trends on resilience to climate change appear similar in both cases, except for indicators related to mortality data and concerning the socio-economic structure.

However, in a more granular analysis, marked differences of urban resilience to climate change in Milan and Rome can be observed across all dimensions and sectors.

Milan stands as a reference model, with high level of urban resilience, especially achieved during the period from 2012 to 2018 (SNPA, 2022). Although the values related to the soil sealing were fluctuating from 2016 to 2020, building construction sites significantly increased in the last two years. As a consequence, the highest value of impermeable surface (188 hectares) was recorded between 2015–2020, being the highest values compared to those observed in the sample of the other Italian cities (Footnote 8). This data can be associated to an increased local vulnerability to landslides and rainfall patterns. Conversely, positive results were achieved in the health sector, especially from 2015–2019, although a worsening was registered after COVID-19 pandemic. During the summer, a positive trend in excess mortality among those over 65 years-population is recorded in Milan (6% lower than expected in 2019). Concerning the percentage of water losses in distribution, values were below the national average, nevertheless, being still high (14.3% in 2018).

---

[10] The Metropolitan city of Milan has 3.2 million residents in 1,575 km^2. Berlin and Barcelona metropolitan area have a similar economic size (Clark et al., 2018).

[11] The Metropolitan City of Rome has a population of 4.3 million residents in 5,363 km^2.

<table>
<tr><th rowspan="2">CITY</th><th colspan="2">LAND AND TERRITORY</th><th>WATER</th><th colspan="2">HEALTH</th><th>GREEN INFRASTRUCTURE</th><th colspan="3">SOCIO-ECONOMIC STRUCTURE</th></tr>
<tr><th>Soil sealing, unused soil (2016-2020) [ha]</th><th>Soil sealing, reversible used soil (2016-2020) [%]</th><th>Total water losses (2012-2018) [%]</th><th>Days of HHWW alert (level 2 and 3) (2015-2019) [n]</th><th>Excess mortality in the summer season (2015-2019) [%]</th><th>Vegetated area on urbanized surface (2016-2020) [%]</th><th>Population < 5 years (2015-2020) [%]</th><th>Population > 65 years (2015-2020) [%]</th><th>Average income per taxpayer (2016-2019) [€]</th></tr>
<tr><td>Milan</td><td rowspan="2">≈</td><td rowspan="2">≈</td><td rowspan="2">↓</td><td rowspan="2">≈</td><td>↓</td><td rowspan="2">=</td><td rowspan="2">↓</td><td>↓</td><td>↑</td></tr>
<tr><td>Rome</td><td>≈</td><td>↑</td><td>≈</td></tr>
</table>

HHWWS: Heat Health Watch Warning Systems

↑ increasing values with a neagtive trend ↑ increasing values with a positive trend ≈ fluctuating values without a clear trend = stable values

Fig. 3 Milan vs. Rome: Resilience to climate change, adapted from SNPA (2022: 34, 74) (Authors' own elaboration).

Regarding the socio-demographic indicators, the decline in births and the increased population aging[12] are typically registered at a national scale. If we consider the increasing migration flows and social inequalities from the central districts to the outskirts (both indicators are not mapped in this report), the sectors with vulnerable populations are expanding, necessitating targeted adaptation policies to effectively address climate challenges. Milan had one of the highest average incomes (€35,336) per taxpayer in 2018, indicating a favourable wealth situation for the population. This suggests a potential correlation with the enhanced capacity of population to access services and information.

Rome presents fluctuations of the performance indicators over time, with a positive trend related to the water use. Similar to Milan, the performance of land and territory indicators in Rome is not satisfactory. Land take, soil consumption, and soil sealing ranked in Rome are among the highest throughout the entire period in the sample.[13] Compared to the vegetated area within the urban belt, being both publicly and privately owned, Rome records a stable value of 56%. In more detail, the total water losses are still higher in Rome when compared to the national data (ranging from 38%–48%), but a positive decrease of water losses was registered in 2018, as a consequence of public investment works on the water supply network.

Factors contributing to one of the main relevant problems in the capital, the emergence of anthropogenic sinkholes, include the faults, leaks, and overall dysfunction in the hydraulic underground network. It is worth noting that an unbroken urban fabric was built over these underground galleries, mostly during the rapid construction boom of the 1950s–1970s, often without thorough technical investigations. Buildings have been often erected directly above shallow underground networks, posing potential risks.

Additionally, a large vulnerable community lives at the mouth of the Tiber River. Over 250,000 residents inhabit areas characterized by a high hydraulic risk, representing one of the highest statistics in Europe (Comune di Roma, 2018). The Tiber River is an important yet still an undervalued blue and green infrastructure for the city and the national territory.

Rome exhibited lower income per taxpayer (€27,766 in 2018) although substantial wealth, surpassing the benchmark set by the sampled municipalities in Italy.

To sum up, the current scenario is marked by positive trends overlaid, however, with specific criticalities, for example related to the heat resilience and water planning, mostly in Rome.

3.1.2 100 Resilience cities' initiative at glance

Programmes like 100 Resilient Cities (100RC) are supporting urban areas in enhancing their disaster preparedness, offering technical support and resources

---

[12] The population aged 65 years and older decreases from 23.9% (2016) to 23% (2020) in Milan (contrary to the national data), and increases from 22.1% (2016) to 22.8% (2020) in Rome.

[13] The Metropolitan City of Rome has the largest consumed area in 2022 with over 70,300 hectares (SNPA, 2023: 85).

to assist cities in formulating and implementing resilience strategies.[14] Promoted in 2013 and concluded in 2019 by an American-based private foundation – The Rockefeller Foundation—100RC network includes six continents, 47 countries, and 16 European cities. Rome joined the 100RC initiative in 2013, whereas Milan took a part of the initiative in 2015.[15] The 100RC was followed by two programmes, Resilient City Catalyst and the new Global Resilient Cities Network (GRCN, which includes the members of the former 100RC programme).[16] It is a city-led organization, whose primary mission is to safeguard vulnerable communities from a range of urban challenges, including those stemming from climate change, as well as physical, social, and economic adversities.

In the 100RC programme, 'shocks' are considered as single event disasters, while 'stresses' as frequently recurring events. The most common shocks faced by the broad 100RC network include rainfall flooding, infrastructure failures, earthquakes and extreme heat. Additionally, common stresses involve aging infrastructure, a shortage of affordable housing, insufficient public transportation, and low social cohesion. These challenges are further exacerbated by the impacts of climate change, evolving global economic dynamics, and increasing inequality worldwide. Rainfall flooding, infrastructure failure, financial and economic crisis, cyber-attacks, and hazardous material accidents are rather common in Milan. Furthermore, the most relevant stresses in Milan are water and air pollution, extreme heat and heatwaves, degradation of urban areas and disruption due to construction sites, as well as social exclusion and inequality, and security risk, influx of migrants, riots or civil unrest, lack of affordable housing, and aging population.

When considering Rome, the most relevant stresses are related to the lack of integrated and updated planning, lack of communication and information among municipalities, slowness and inefficiency of bureaucracy, limited access to broadband telecommunications, pollution of air, water, and soil, critical issues in the urban solid waste cycle, loss of ecosystem services, deterioration of some public areas and buildings, lack of civic sense, and aging infrastructure (Comune

---

[14] Cities in the 100RC network (https://www.100resilientcities.org/) receive direct funds to create a resilience roadmap, i.e., (i) financial and logistical guidance for appointing a Chief Resilience Officer (CRO); (ii) expert support for defining a strong strategy; (iii) access to solutions and partners from various sectors for strategy development; and (iv) membership in a global network of cities for mutual learning and assistance. The city resilience framework devised within 100RC includes several dimensions, i.e., health and well-being, economy and society, infrastructure and environment, and leadership and strategy. For more information: Coppola et al. (2020); Galderisi et al. (2020).

[15] Galderisi et al. (2020) stress the delays in both processes, for instance, the Chief Resilience Officer (CRO) was defined in Milan in 2018, three years after its inclusion in 100RC. In Rome, the city partner (from ICLEI to ARUP) was changed at the end of 2015. Coppola et al. (2020) comparatively analyse Milan and Rome considering the downscaling, internal institutionalization, and external institutionalization.

[16] See Resilient City Catalyst (https://www.rcc.city/) and new Global Resilient Cities Network (https://www.rockpa.org/project/global-resilient-cities-network/).

di Roma, 2018). The most common shocks in Rome are earthquakes, sinkholes, landslides, flash and river floods, and drainage flooding.

Milan and Rome are rapidly growing urban areas with high population density, thus marked by a certain complexity and uncertainty (Coppola et al., 2020; Coppola and Haupt, 2022). Both have a "mayor-centred local political system, limited fiscal autonomy, and strong dependence on state money transfers" (Coppola et al., 2020: 9). Both cities are characterized by weak administration systems in terms of human and economic resources, yet they diverge in terms of political and economic systems. Milan is more politically stable, with a prevalence of centre-left party, experiencing significant socio-economic growth and projecting an international image through the mayor and the whole system (including several international events, such as Expo 2015). Rome was alternately governed by different political parties, featuring low networked governance, urban crisis, corruption, and chronic services deficiencies. In contrast, Milan exhibited a more robust track record and continuity of policy activism, notably through the establishment of science-policy collaborations focused on climate adaptation, involving both local (metropolitan area) and academic (Politecnico di Milano) institutions.[17] Conversely, Rome displays a weaker pattern in recent environmental and energy plans, with limited science-policy collaborations and no formal partnerships. The city lacks significant engagement on resilience involving various actors such as enterprises and philanthropies.

Focusing on the measures to mitigate emerging challenges and reach climate neutrality by 2050, Milan needs to address the urban heat island and energy poverty challenge. For this reason, as one of several flagship projects, Milan has been developing the *ForestaMI*, an *Urban Forest strategy*,[18] aimed to increase the tree cover ratio to combat urban heat island impact, improve air and water quality, reduce the risk of rainfall flooding, and reduce the decay of urban areas and spaces. Societal inequality is also a pressing problem, and a €300-million Peripheries Plan was launched for addressing this challenge. Combined with the investment in inner suburbs, other projects have been implemented to respond to the need of extra residential and commercial spaces (e.g., projects such as *Porta Nuova, City Life*, *Milanosesto*, and *Milano Santa Giulia*) (Clark et al., 2018).

In Rome, one of the action priorities involves the management of the Tiber River basin, through the implementation of projects coordinated by the Special Tiber Office. The masterplan aims to recover the relationship between the city and its main watercourse through new waterfront parks, sports infrastructures, and the creation of other public spaces to improve the quality of life for the residents.

---

[17] As of today, Milan has been involved in nine transnational municipal networks (TMNs) and 16 EU-project-based networks (e.g., Horizon, 2020), while Rome has been involved in two TMNs and five EU-project-based networks (see Coppola et al., 2020: 11). Milan and Rome are involved in a complementary programme for climate change mitigation actions, C40 (https://www.c40.org/).

[18] https://forestami.org/en/il-progetto/

3.1.3 Digital Twin technologies under the lens of urban resilience

As deduced from the previous brief overview of resilience framework indicators and current reports on the existing state of resilience levels in Italian cities (e.g., SNPA, 2022), the identification of the critical factors hindering the achievement of resilient goals in Milan and Rome and the respective mitigation measures is a significant challenge. Factors such as political instability, limited internationalization, and the absence or discontinuous collaboration between academia and public entities can significantly impact the effectiveness of urban resilience-oriented strategies (Coppola et al., 2020).

The utilization of digital methods, tools, and information services has proven highly effective in enhancing and accelerating the development of urban resilience. Digital information services, technological innovations, and novel models of urban governance are crucial to get closer to the achievement of EU targets by 2030. Emerging digital environments, based on both desk research and stakeholders' engagement activities, are built upon shared, visualized, processed real-time data streams. Drawn from diverse sources such as grey literature, social media channels, phones, environmental sensors, and smartwatches, these data enhance the management capabilities, facilitating the attainment of resilient targets. The impact is multifaceted, spanning across various phases. In fact, big data analytics in Disaster Risk Management can be predictive (forecasting, simulating, modelling data, assimilation), descriptive (understanding, monitoring, detection, comparison), and prescriptive (nudging behaviour analysis sensitization) (Vermiglio et al., 2020). Other innovative digital architectural approaches and planning tools in near-real time simulated scenarios include Geographic Information System (GIS)-based and point-cloud modelling techniques, the application of Life Cycle Assessment (LCA), building information models (BIM), and heritage building information models (HBIM). Applied to the city, the computer-modelled environment is defined as City Digital Twin (CDT), as stated by Therias and Rafiee (2023: 4171): "CDT is the bi-directional multi-layered digital representation of a city that enables the visualization, simulation, and evaluation of urban interventions based on dynamic data about the physical and social systems that are embedded in the city itself."

At the architectural scale, Digital Building Renovation Passport (BRP) tool is a long-term (15–20 years) renovation plan available in electronic or paper format. It guides deep renovations, ensuring seamless coordination and offering personalized recommendations for building owners or building residents, expediting the renovation process. It can be linked to a logbook, serving as a digital repository for regularly updated building information.

4. Limitations and Challenges

While digital applications present significant opportunities across various sectors, diverse concerns have been highlighted by scholars and practitioners (e.g., Lei et al., 2023). Addressing the lack of public participation and transparency in digital twin data, enhancing the accuracy and trustworthiness of data models, and improving data representativeness are all crucial challenges that must be tackled to prevent the deployment of de-biasing data services. Concerns about the representativeness and quality of the data collected and processed are tied to the issues of low social equality, as digital information systems and services can potentially widen the digital gap. In the design of new technologies for climate change, a paramount objective should be to mitigate social inequalities stemming from the often overlooked contributions of women and girls,[19] address the pressing needs of ageing population and low-income people.

Furthermore, individual privacy protection risks are often framed as relevant obstacles of environmental policies. Beside legal and ethical issues, consideration must be given to how to update the information infrastructure and to educate skilled workload. Operationally, leveraging rich and heterogeneous datasets, along with tasks such as operability (e.g., converting BIM data to GIS), involves complex phases that demand highly skilled teams and equipments. The use of these digital twins' technologies are still mockups, used in research stages and inadequately implemented in real case studies.

In Italy, it is worth noting that there is often a slower uptake and lower familiarity with digital tools compared to other European countries (Italia, 2021), and this trend is also evident in Rome (Comune di Roma, 2018). In the *National Recovery and Resilience Plan* (PNNR in italian), emphasis is placed on the importance of fostering digital citizenship (Italia, 2021). Both the private and public sectors are marked by a lack of familiarity with digital technologies. A considerable segment of the population in Italy currently lacks these skills, impeding their active engagement in democratic processes, limiting access to employment opportunities, and hindering the adoption of digital services. Limited digital skills hinder the execution of emergency plans, especially when dealing with vulnerable households and communities to disasters, including older people and low-income and low-literate inhabitants, war refugees, as well as individuals who are temporarily displaced from their residences due to natural disasters. It is thus essential to raise awareness about climate adaptation and the issuance of early warnings.

5. Climate Action Pathways in Milan and Rome

To improve the resilience level in Milan and Rome, we can underline, without claiming to be exhaustive, climate action pathways considering four interconnected spheres of priority actions, as follows:

- ***Social dimension:*** Digital approaches and tools can directly contribute to a behavioural change of people for the reduction of consumption patterns. Digital hubs and platforms can especially impact the social sector, since people need to access to service and citizens' network must be active (Figueiredo et al., 2018: 18). Communities should actively address local challenges, sharing insights

[19] See on climate injustice: Fisher, 2015; environmental injustice: Chakraborty, 2017.

and evolving needs with municipal representatives for a more informed and responsive approach to their unique dynamics. If we focus on communication (see Figure 2), the improvement of local interaction and sociability levels is fundamental to minimize negative impacts of disasters and their ripple effects. In Rome, the *Digital Agenda of Rome Capital* outlines indicators of a prospective enhancement in the communication between citizens and the government, achieved through the integration of digital services and processes. In Milan, a relevant demonstration of citizen engagement unfolded through the stakeholder mapping process, using e-tools is the Air and Climate Plan for the period 2020–2030.[20] This e-government initiative involves the critical analysis of multiple-choice questionnaire, the organization of thematic laboratories and district meetings, and other participatory planning activities, whose inputs are integrated into the current city council plans.

- ***Economic dimension:*** Digital twins serve as a replica that highlights the societal benefits arising from the funding investment. As defined in the report published by Arup (2019), digital twins can lay the groundwork for improving social equity. This entails combining open access to system information with crowdsourcing platforms for fundraising to promote equity and differentiate investments. For example, in Milan, the *Civic Crowdfunding Initiative* is an online platform enabling citizens to fund projects with a social impact.[21] In 2016, Milan launched *Digital Bridges*, an international learning initiative designed for city governments to exchange best practices among the most innovative cities worldwide.
- ***Environmental dimension:*** Urban heat island and water-related issues can be also addressed using digital twins technology, such as vulnerability simulators[22] and flood forecasting based on machine learning (ML), i.e., hydraulic models, system-level models, Monte Carlo analysis,[23] smart networks (Arup, 2019). ML methods enables the processing of a wide range of remote sensing datasets, e.g., using UAV (Unmanned Aerial Vehicles) or satellite imagery. As stressed by Therias and Rafiee (2023), the Extreme Weather Layer combines the digital twin of the urban drainage system (UDS) with the GIS to help city governments (among others) analyse the effects of climate change and new developments. GIS-based tools enable the assessment of the risk of flooding at both individual plots and the urban scale on existing infrastructure. This involves using data like UDS information, rainfall patterns, official weather data, land use, and topography. Hydraulic modelling software is then used to identify vulnerable manholes in the urban area, susceptible to heavy rainfall and street flooding (Truu et al., 2021).

---

[20] https://www.comune.milano.it/aree-tematiche/ambiente/aria-e-clima/piano-aria-clima.
[21] https://economiaelavoro.comune.milano.it/progetti/ crowdfunding-civico-seconda-edizione.
[22] e.g., https://www.siradel.com/urban-climate-resilience-toolkit/
[23] This method was applied to the Tiber River in the research conducted by Natale and Savi (2007).

- ***Institutional dimension:*** The adoption of digital twin technologies offers relevant opportunities to enhance real-time monitoring, improve transparency regarding priority actions, and streamline informed decision-making processes. Collaborating with governmental authorities, these technologies can facilitate the implementation of complementary activities. Public authorities, city administrators, and civil society organizations must collaborate to establish a cohesive network and implement effective climate change initiative. In Milan, current strategies are directed to improve the collaboration at city-region levels, and with Turin and the Alpine region (Clark et al., 2018). In Rome, collaborative efforts should be enhanced following the activities undertaken by the technical committee to safeguard and valorize a priority intervention area along the Tiber delta, involving the Lazio Region, the Basin Authority of the Central Apennines Hydrographic District, the Harbor Master's Office of Rome, the Port Authority of Civitavecchia Fiumicino and Gaeta, and the Tiber and Roman Agro Reclamation Consortium (Comune di Roma, 2018).

6. Final Considerations

Resilient cities should be robust, redundant, flexible, resourceful, reflective, inclusive, and integrated systems. Effective urban resilient policies must prioritize the valorization of context-based features of the urban environment and community assets. Local urban policies should establish a cross-sectorial and cross-scale plan that considers uncertainties (unpredictability of the event) and embraces the evolutionary paradigm of the resilience. Facilitating the green transition of future cities can be achieved through multi-stakeholder engagement and collaborative networks that leverage the potential of digital information services. Within a broad urban resilience framework based on a whole-of-society and city-centric approach, this study focuses on the local resilience levels in two Italian cities, Milan and Rome. By comparing the information provided by the *Sistema Nazionale per la Protezione dell'Ambiente* (SNPA, 2022) and analysing studies on the 100RC initiative (e.g., Coppola et al., 2020; Galderisi et al., 2020), a disparity in efforts to meet resilient goals between Milan and Rome can be observed. When examining these two case studies, it is evident that resilience thinking strategies must be developed considering both political and functional aspects, physical and intangible assets.

Given the intricate nature of urban systems like Milan and Rome, characterized by their complexity and sectoral interconnections, it is crucial to develop tailored digital tools and integrated approaches that address the specific needs of each local context and align with various spheres of priority actions.

Acknowledgements

This work was partially supported by the *Fundação para a Ciência e a Tecnologia* through ISTAR Projects with grant numbers UIDB/04466/2020 and UIDP/04466/2020.

References

Albrechts, L., Barbanente, A. and Monno, V. (2020). Practicing transformative planning: The territory-landscape plan as a catalyst for change. *City, Territory, and Architecture, 7*(1), 2–13. https ://doi.org/10.1186/s4041 0-019-0111-2.

Alexander, D.E. (2013). Resilience and disaster risk reduction: An etymological journey. *Natural Hazards and Earth System Sciences, 13*, 2707–2716. https://doi.org/10.5194/nhess-13-2707-2013.

Arup (2019). *Digital Twin: Towards a Meaningful Framework.* Available: https://www.arup.com/perspectives/publications/research/section/digital-twin-towards-a-meaningful-framework.

Bahadur, A. and Tanner, T. (2014). Transformational resilience thinking: Putting people, power, and politics at the heart of urban climate resilience. *Environment and Urbanisation, 26*(4), 1–15. doi:10.1177/0956247814522154.

Baravikova, A., Coppola, A. and Terenzi, A. (2020). Operationalizing urban resilience: Insights from the science-policy interface in the European Union. *European Planning Studies, 29*(2), 241–58. https://doi.org/10.1080/09654313.2020.1729346.

Carpenter, S.R., Westley, F. and Turner, G. (2005). Surrogates for resilience of social–ecological systems. *Ecosystems, 8*, 941–44. https://doi.org/10.1007/s10021-005-0170-y. 100 Resilient Cities. Available: https://www.100resilientcities.org/

Chakraborty, J. (2017). Focus on environmental justice: New directions in international research. *Environmental Research Letters, 12* (3). 030201. doi:10.1088/1748-9326/aa63ff.

Clark, C., Moonen, T. and Nunley, J. (2018). *Milan's Competitiveness. Milan Case Study*. Urban Land Institute.

Coaffee, J. et al. (2018). Urban resilience implementation: A policy challenge and research agenda for the 21st century. *Journal of Contingencies and Crisis Management, 26*, 403–410. https://doi.org/10.1111/1468-5973.12233.

Comfort, L.K., Sungu, Y., Johnson, D. and Dunn, M. (2001). Complex systems in crisis: Anticipation and resilience in dynamic environments. *Journal of Contingencies and Crisis Management, 9*, 144–58.

Comune di Milano. (n.d.). *Augmented (Human) Capital Innovation through Resilience, Response, and Renewal in Milano*. https://www.comune.milano.it/documents/20126/272176777/Dossier+iCapital+Milan.pdf/2359a4df-2cb0-e2f8-15f7-780d3154213b?t=1627975738508.

Comune di Roma. (2018). *Roma Strategia di Resilienza*. https://www.comune.roma.it/web-resources/cms/documents/strategiaresilienza180618.pdf.

Coppola, A., Crivello, S. and Haup, W. (2020). Urban Resilience as New Ways of Governing: The Implementation of the 100 Resilient Cities Initiative in Rome and Milan. *In*: A. Balducci, A., D. Chiffi and F. Curci, (Eds.), *Risk and Resilience.* SpringerBriefs in Applied Sciences and Technology.

Coppola, A. and Haupt, W. (2022). Philanthropic Organisations and the Global Circulation of Urban Resilience Practices: The Case of 100 Resilient Cities. *In*: *The Palgrave Handbook of Global Sustainability*. Cham. https://doi.org/10.1007/978-3-030-38948-2_167-1.

Cutter, S.L., Ash, K.D. and Emrich, C.T. (2014). The geographies of community disaster resilience. *Global Environmental Change, 29*, 65–77. https://doi.org/10.1016/j.gloenvcha.2014.08.005.

Davoudi, S. (2012). Resilience: A Bridging Concept or a Dead End? *Planning Theory and Practice, 2*(13), 299–307. https://doi.org/10.1080/14649357.2012.677124.

Davoudi, S., Shaw, K., Haider, L.J., Quinlan, A.E., Peterson, G.D., Wilkinson, C., Fünfgeld, H., McEvoy, D. and Porter, L. (2012). Resilience: A bridging concept or a deadend? "Rreframing" resilience: Challenges for planning theory and practice interacting traps: Resilience assessment of a pasture management system in northern Afghanistan urban resilience. *Plan. Theory Pract., 13*(2), 299–333.

Desouza, K.C. and Flanery, T.H. (2013). Designing, planning, and managing resilient cities: A conceptual framework. *Cities*, 89–99. https://doi.org/10.1016/j.cities.2013.06.003.

Figueiredo, L., Honiden, T. and Schumann, A. (2018). *Indicators for Resilient Cities*. OECD Publishing. https://dx.doi.org/10.1787/6f1f6065-en.

Fisher, S. (2015). The emerging geographies of climate justice. *The Geographical Journal, 181*(1), 73–82. http://www.jstor.org/stable/43868629.

Galderisi, A., Limongi, G. and Salata, K.D. (2020). Strengths and weaknesses of the 100 Resilient Cities Initiative in Southern Europe: Rome and Athens' experiences. *City, Territory, and Architecture, 7* (16). https://doi.org/10.1186/s40410-020-00123-w.

Holling, C.S. (1973). Resilience and Stability of Ecological Systems. *Annual Review of Ecology and Systematics, 4*, 1–23. https://doi.org/10.1146/annurev.es.04.110173.000245.

Holling, C.S., Gunderson, L. and Lance, H. (2002). Resilience and Adaptive Cycles. *In*: L.H. Gunderson and C.S. Holling (Eds.), *Panarchy: Understanding Transformations in Human and Natural Systems*, 25–62, Island Press, Washington D.C..

Intergovernmental Panel on Climate Change (IPCC). (2023). *Climate Change 2022: Impacts, Adaptation and Vulnerability*. Working Group II Contribution to the Sixth Assessment Report of the Intergovernmental Panel on Climate Change. Cambridge. doi:10.1017/9781009325844.

Italia. (2021). *Piano Nazionale di Ripresa e Resilienza* Available from: https://www.governo.it/sites/governo.it/files/PNRR.pdf.

Lei, B., Janssen, P., Stoter , J. & Biljecki, F. (2023). Challenges of urban digital twins: A systematic review and a Delphi expert survey, *Automation in Construction, 147*, 104716. https://doi.org/10.1016/j.autcon.2022.104716.

Manca, A.R., Benczur, P. and Giovannini, E. (2017). Building a scientific narrative towards a more resilient EU society, Part 1: A conceptual framework, EUR 28548 EN. doi:10.2760/635528.

Manyena, S.B. (2006). The concept of resilience revisited, *Disasters, 30*, 434–50.

Natale, L., Savi, F. (2007). Monte Carlo analysis of probability of inundation of Rome. *Environmental Modelling & Software, 22* (10), 1409–16. https://doi.org/10.1016/j.envsoft.2006.12.004.

Park, A. (2011). Beware paradigm creep and buzzword mutation, *Forest. Chron., 87*, 337–44. https://doi.org/10.5558/tfc2011-023.

Peters, K. et al. (2016). *'Resilience' across the Post-2015 Frameworks: Towards Coherence*? ODI Working Paper, Overseas Development Institute, London.

Reghezza-Zitt, M., Rufat, S., Djament-Tran, G., Le Blanc, A. and L'Homme, S. (2012). What resilience is not: Uses and abuses. *CyberGeo, 2012*(621), 1–23.

Resilience Alliance. (2007). *Urban Resilience Research Prospectus*. A Resilience Alliance Initiative for Transitioning Urban Systems towards Sustainable Futures, CSIRO/Arizona State University/Stockholm University, Australia/USA/Sweden.

Saja, A.M.A., Goonetilleke, A., Teo, M. and Ziyath, A.M. (2019). A critical review of social resilience assessment frameworks in disaster management, *International Journal of Disaster Risk Reduction*. https://doi.org/10.1016/j.ijdrr.2019.101096.

Sistema Nazionale per la Protezione dell'Ambiente (SNPA). (2022). *Città in transizione: i capoluoghi italiani verso la sostenibilità ambientale. Documento di sintesi. [Cities in Transition:Italian Regional Capitals toward Environmental Sustainability. Summary Document] Report SNPA* 31/2022.

Sistema Nazionale per la Protezione dell'Ambiente (SNPA) (2023). *Consumo di suolo, dinamiche territoriali e servizi ecosistemici – Edizione 2023. [Land Consumption, Territorial Dynamics, and Ecosystem Services – 2023 Edition] Report*. SNPA 37/2023.

Therias, A. & Rafiee, A. (2023). City digital twins for urban resilience. *International Journal of Digital Earth, 16*(2), 4164–90, DOI: 10.1080/17538947.2023.2264827.

The Rockefeller Foundation & Arup. (2015). *City Resilience Framework*. https://www.rocke felle rfoun datio n.org/wp-conte nt/uploa ds/City-Resilience-Frame work-2015.pdf.

The Rockefeller Foundation & Arup. (2016). *City Resilience Index: Understanding and Measuring City Resilience*. http://publications.arup.com/publications/c/city_resilience_index.

Truu, M., Ivar, A., Roosimägi, J., Kändler, N., Vassiljev, A. and Kaur, K. (2021). Integrated Decision Support System for Pluvial Flood-Resilient Spatial Planning in Urban Areas. *Water, 13* (23): 3340. https://doi.org/10.3390/w13233340.

UN (2015). Transforming Our World: *The 2030 Agenda for Sustainable Development*. A/RES/70/1, United Nations. sustainabledevelopment.un.org.

UN (2019). *The Sustainable Development Goals Report 2019*. United Nations, New York. https :// www.unsta ts.un.org/sdgs/repor t/2019/The-Sustainable-Development-Goals -Report-2019.pdf.

UN-Habitat (2017). *Trends in Urban Resilience, Nairobi: UN-Habitat*. ISBN: 978-92-1-132743-4.

UN-Habitat (2016a). *New Urban Agenda*, A/CONF.226/4, adopted in Quito, Ecuador, 17–20 October 2016. *In:United Nations Conference on Housing and Sustainable Urban Development*. http:// habitat3.org/thenew-urban-agenda.

UN-Habitat (2016b). SDG Goal 11 Monitoring Framework: A guide to assist national and local governments to monitor and report on SDG Goal 11 indicators, March. *In*: *United Nations Conference on Housing and Sustainable Urban Development*. https://unhabitat.org/sdg-goal-11-monitoringframework.

United Nations Office for Disaster Risk Reduction (UNISDR). (2012). *Making Cities Resilient Report 2012: My City is Getting Ready*. United Nations Office for Disaster Risk Reduction. www.unisdr. org/files/28240_rcreport.pdf.

Vermiglio, C., Kudo, H. and Zarone, V. (2020). *Making a Step Forward towards Urban Resilience. The Contribution of Digital Innovation*. doi:10.1007/978-3-030-52869-0_10.

Wilkinson, C. (2011). Social-ecological resilience: Insights and issues for planning theory. *Planning Theory*, *2*, 148–69.

Wisner, B. and Adams, J. (2002). *Environmental Health in Emergencies and Disasters: A Pratical Guide*. World Health Organization.

11 Current Trends and Challenges in BIM–LCA Integration

Sara Parece,[1*] Ricardo Resende[2] and Vasco Rato[3]

1. Introduction

The Architecture, Engineering, Construction, and Operation (AECO) industry faces challenges that threaten its long-term sustainability, such as high greenhouse gas emissions and energy consumption, low productivity, and lack of technological evolution. The built environment accounts for 30–40% of global carbon dioxide emissions and 40% of global energy consumption (United Nations Environment Programme, 2024). European initiatives such as the Green Deal and the Renovation Wave emphasise reducing carbon dioxide emissions and achieving carbon neutrality by 2050 (Renovation Wave, 2024). These initiatives call for extensive building renovation and rehabilitation, effective measures to reduce embodied and operational carbon footprints, and increased data digitalization throughout the lifecycle of buildings (Röck et al., 2020).

Life Cycle Assessment (LCA) is a widely accepted method for evaluating a building's environmental footprint throughout its life cycle. The ISO 14040 standard (ISO-14040, 2006) establishes a formal process for conducting an LCA of industrial processes. Unlike other industrial processes, which typically have limited elements with short lifespans and standard features, buildings have a wide range of elements with long lifespans and unique characteristics, so several LCA frameworks compatible with the construction industry have been proposed by EN 15978: 2011 and EN 15804: 2012. Building LCA involves multiplying all products, materials, and processes, which are usually compiled in a bill of quantities (BoQ), by environmental data cost from generic LCA databases or environmental product declarations (EPDs) (Cavalliere et al., 2019).

BIM (Building Informaion Modelling) digitally represents the physical and functional characteristics and information of construction projects, thereby simplifying LCA's data acquisition (Cheung et al., 2012). Detailed quantities can be extracted automatically, which speeds up the LCA calculations.

[1,3] ISTAR-Information Sciences and Technologies and Architecture Research Center, ISCTE – University Institute of Lisbon, School of Technology and Architecture.
[2] ISTAR-Information Sciences and Technologies and Architecture Research Center, ISCTE – University Institute of Lisbon, Department of Applied Digital Technologies.
Email: ricardo.resende@iscte-iul.pt[2]; vasco.rato@iscte-iul.pt[3]
* Corresponding author: sara_parece@iscte-iul.pt

Numerous BIM-LCA integrations have been proposed in recent years. Still, there is no agreement on which approach is the most effective or accurate since this depends on the design stage, the maturity of the information in BIM models, the BIM workflow implemented in the design company, and the overall goal of the LCA. Additionally, there is a gap between industry needs and academic research. Although studies have addressed the integrated use of BIM and LCA, they fail to identify the current stage of BIM-LCA technology and the main literature gaps.

This paper contributes to a systematic understanding of BIM-LCA integration identifying the current trends and challenges. The review covers academic articles on BIM-based LCA and commercial BIM-based LCA tools. The methodology for acquiring and categorizing the literature is detailed in Section 2. Section 3 summarizes and discusses the findings from the systematic literature review, while Section 4 presents the main conclusions.

2. Research Methods

Following the guidelines of the Preferred Reporting Items for Systematic Reviews and Meta-Analysis (PRISMA) (Page et al., 2021), a systematic review was employed to explore BIM-based LCA applied in new buildings and renovation projects. This study aims to answer the research question: *What are the limitations, trends, and future developments of BIM-based LCA for building design?* The methodology involved systematic searching, screening, selection, and comprehensive reporting of pertinent literature. Two search strings were selected, targeting article titles, abstracts, and keywords across two journal databases, Scopus and Web of Science (WoS) (Table 1). Only peer-reviewed journal articles in English from 2013–March 2024 were included, as they generally provide complete and high-quality information, while review articles, editorials, and conference papers were excluded. The search initially identified 749 publications. After removing 288 duplicates and applying the inclusion and exclusion criteria, 65 articles were selected (Figure 1).

Table 1 Search strings and results obtained.

Search strings	*Journal Database*	*Frequency of hits*	*Added to screening*
("life cycle assessment" OR "life cycle environment* assessment" OR "life cycle analysis" OR "life cycle environment* analysis" OR "lifecycle assessment" OR "lifecycle analysis" OR LCA) AND ("building information model*" OR "BIM")	Scopus	309	309
	WoS	252	86
(GWP OR "Global Warming Potential" OR "Carbon Footprint" OR "Embodied Carbon") AND ("building information model*" OR "BIM")	Scopus	108	48
	WoS	80	18
Total		749	461

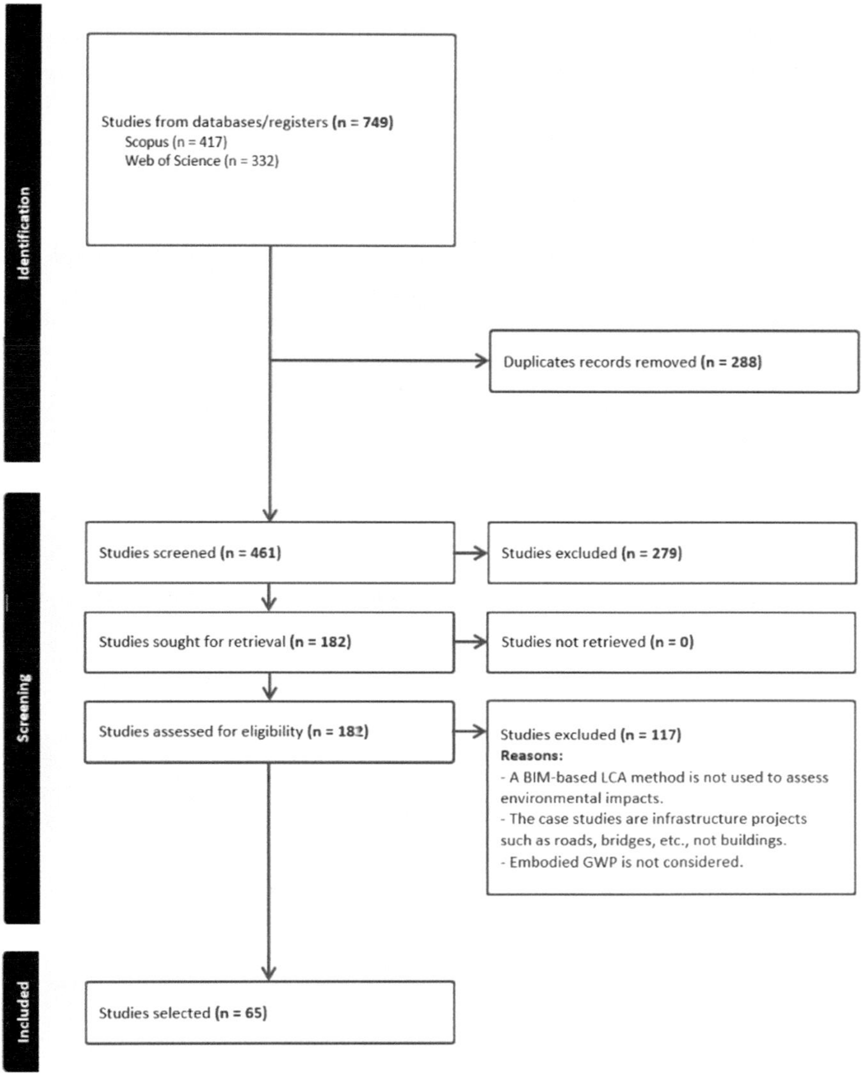

Fig. 1 Research framework

The screening process and selection of the 65 articles were guided by the following criteria:

- The article employs a BIM-based LCA integration to evaluate the environmental impact of new building construction, renovation, or refurbishment projects.
- The focus is on buildings, excluding infrastructure projects such as roads, bridges, and railways.
- The life-cycle concept includes considerations of Global Warming Potential (GWP) related to materials, also known as embodied carbon.

Furthermore, we aimed to highlight articles that addressed GWP and indicators such as life-cycle cost (LCC), design for disassembly (DfD), and design for adaptability (DfA), and articles that applied multi-criteria analysis (MCA) or multi-criteria decision analysis (MCDA), optimisation techniques, and machine learning methods to improve decision-making.

3. Results and Discussion

In recent years, many articles on LCA integrated into BIM have been published with a growing trend. This field of research has been the focus of attention and rapid developments in particular over the last five years. Figure 2 demonstrates the 65 selected publications by year, from 2013 to March 2024.

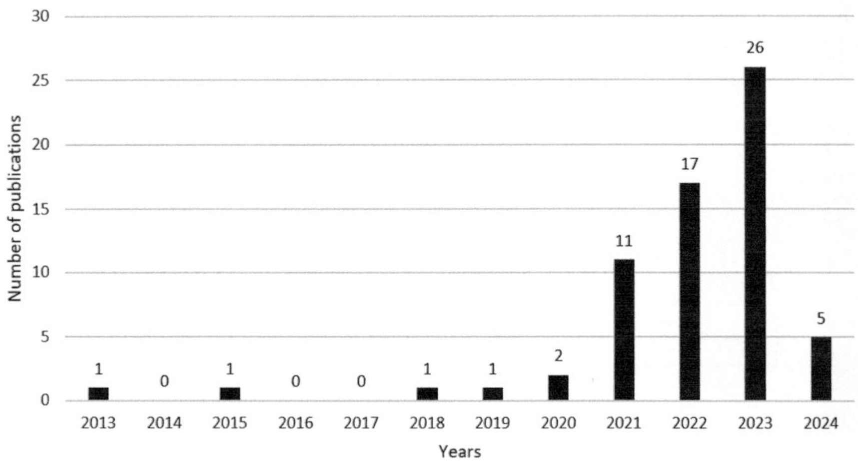

Fig. 2 Publications selected for full-text analysis by year.

Scientometric analysis was conducted using the VOSviewer software. VOSviewer is a tool for constructing and visualizing bibliometric networks (VOSviewer: Visualizing Scientific Landscapes, n.d.). Keywords serve as representative descriptions and directions of the research article's content. The minimum keyword frequency in VOSviewer was set to two. A filtering process followed, which involved merging synonyms (e.g., "carbon footprint" and "carbon emissions"), expanding abbreviated terms into their full forms (e.g., "BIM" to "building information modelling"), and excluding general terms. Twenty keywords were selected, as shown in Table 2 and Figure 3. A large node size signifies high frequency, a thick connection line denotes a strong relationship, and different colours categorise nodes into clusters. The results identified the five most prominent topics in the knowledge area of BIM-based LCA as life-cycle assessment, building information modelling, environmental impacts, global warming potential, and sustainability. Figure 3 demonstrates a strong correlation between LCA and BIM keywords, indicating that BIM-based LCA is extensively utilised in the literature.

Table 2 Details of the keywords of the selected articles.

Keywords	*Occurrences*	*Total Link Strength*
building information modelling	42	93
life-cycle assessment	51	63
environmental impact	7	21
sustainability	7	21
global warming potential	7	19
life cycle cost	6	13
carbon footprint	5	8
embodied environmental impact	4	13
buildings	3	6
embodied carbon	3	9
energy consumption	3	1
level of development (loD)	3	9
decision-making	2	4
design process	2	8
embodied energy	2	4
multi-objective optimisation	2	5
optimisation	2	7
RC structures	2	7
renovation	2	6
residential building	2	6

3.1 BIM Use

A BIM digitally represents buildings' physical and functional attributes. BIM offers numerous advantages, including enhanced visualisations, integration and relationship between data and geometry and, stakeholders' coordination. It is an valuable tool that significantly reduces the time and effort needed to manage building data. In the context of LCA, BIM enables the automated extraction of detailed quantities from models, facilitating the compilation of the life cycle inventory. Autodesk Revit (Architecture, Structures, and MEP) is the most common software for BIM-based LCA, followed by Graphisoft ArchiCAD, as illustrated in Figure 4(a).

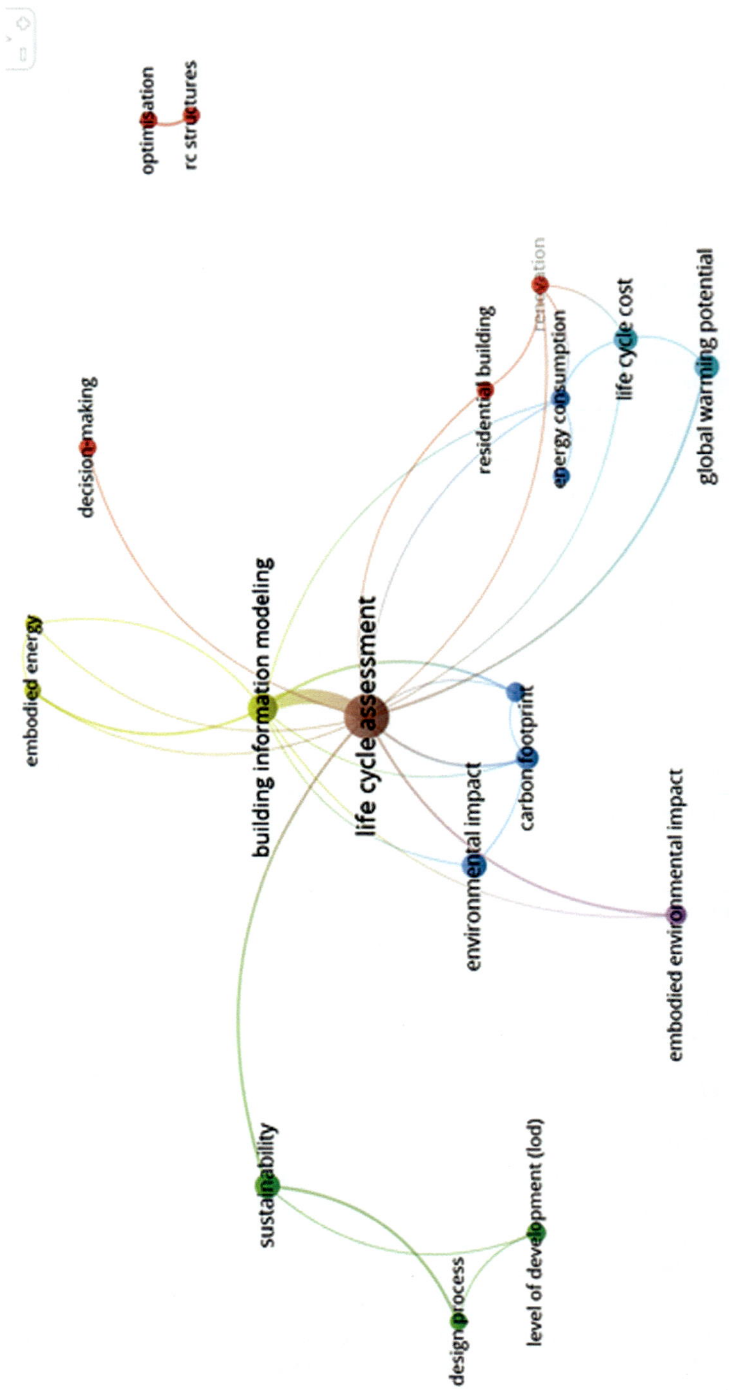

Fig. 3 Network of co-occurring keywords (Authors' own elaboration).

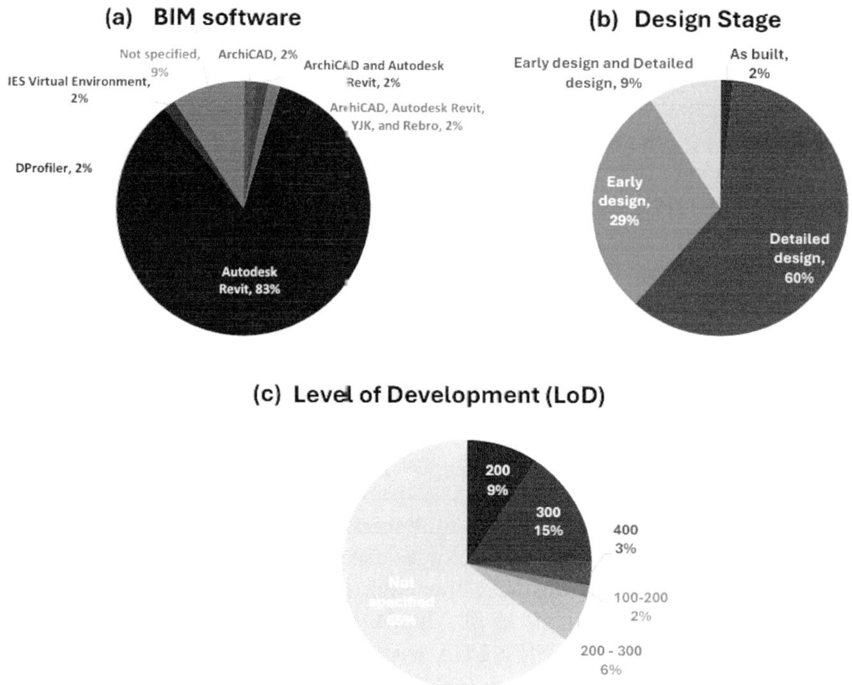

Fig. 4 Charts with the 65 case studies categorized by BIM software, design stage, and LoD.

The materials and components of a building are represented in a BIM model, both geometrically and non-geometrically. Their representation varies based on the Level of Development (LoD), defined by the American Institute of Architects as "the minimum dimensional, spatial, quantitative, qualitative, and other data included in a Model Element." LoD is categorized into five levels, ranging from 100 to 500. Although the term "Level of Development" has been largely replaced by "Level of Information Need" (LoIN), the majority of the articles analyzed still reference LoD (AIA, 2013).

Conducting an Building LCA in the early stages of the project is challenging. At this point of the design, it is inefficient to model with a high LoD, so designers rely on simplifications or adaptations in geometry and information. If these are not applied correctly, they can compromise the accuracy of the QtO and LCA results, potentially leading to incorrect decisions. Nonetheless, it is during this phase that project decisions are most impactful and least expensive to modify.

Additionally, we must consider heterogeneity in model creation, as models are not developed simultaneously across all specialities. As a general rule, LoD increases gradually to meet the demands of each design phase, but that is not always the case. For example, structural elements or external wall composition can be detailed early in the process, whereas interior finishes only later during execution. In this study, we follow the division of the building design process into three stages: early design stages, detailed design stages, and construction design stages, as in Sharif and Hammad (2019) and Zhou, Ma, et al. (2023).

While most articles did not specify the LoD (as shown in Figure 4 (c)), 10 articles identified LoD 300 as appropriate for accurate LCC and LCA calculations: (Alwan et al., 2021; Arvizu-Piña et al., 2023; Cheng et al., 2022; Kim and Kim, 2023; Lee et al., 2021; Llatas et al., 2022; Ma et al., 2024; Nehasilová et al., 2022; Su et al., 2021; Zhang et al., 2022). At LoD 300, key materials and components are specified, including the actual sizes, shapes, locations, wall layers, and structural elements. Six works performed a simplified LCA study with LoD 200 BIM models: (Ajayi et al., 2015; Forth, Abualdenien, et al., 2023; Forth, Hollberg, et al., 2023; LLatas et al., 2022; Sobhkhiz et al., 2021; Soust-Verdaguer, Gutiérrez Moreno, et al., 2023).

Furthermore, 19 articles focus on early design LCA studies, while 39 aim for highly refined design phases (as shown in Figure 4 (b). The use of different tools for each design phase hinders a continuous and accurate LCA. Additionally, using various databases and methodologies can significantly alter the results. We have identified a lack of tools that allow environmental impacts to be monitored throughout the project (Bueno et al., 2018). As suggested by Cavalliere et al. (2019) these tools should utilise LCA databases ranging from generic data to Environmental Product Declarations (EPD), aligning the LoD with the design phase. Hollberg et al. (2020) continuously monitored GWP during the design phase and concluded that is often significantly overestimated in the early design stage due to model simplifications. Parece, Resende, et al. (2024) proposed a method adapted to different modelling techniques, LoD and design phases, which involves using a database of possible constructive solutions with maximum and minimum GWP values to complement the absolute quantities of the BIM objects in the model.

3.2 LCA Calculation

The International Standards Organization (ISO) and the CEN/TC 350 committee have developed comprehensive standards to assess environmental lifecycle impacts in construction projects. These standards cover the general framework, principles, calculation methods, lifecycle stages, impact categories and EPDs. While ISO 14040: 2006 is applicable across all sectors, other standards, such as ISO 15392: 2019, ISO 21678: 2020, EN 15978: 2011, and EN 15804: 2012, specifically address the use of LCA in construction activities. Together, these international standards encompass the essential components needed to conduct an LCA study for construction projects.

According to ISO 14040 and ISO 14044, LCA is divided into four steps: (1) Goal and scope definition: defining the goal, scope, and boundary of the LCA research; (2) Life cycle inventory: collecting data on products, materials, processes, and energy consumption to create an input or output inventory; (3) Life cycle impact assessment (LCIA): measuring flows, to determine an overall impact category total; and (4) Interpretation: evaluating the results from the life-cycle inventory and LCIA to draw conclusions and make recommendations.

The extensive range of impact categories in standard LCIA techniques adds complexity to LCA outcomes. Using a single score indicator can simplify decision-making while reducing the relevant information. For example, the CML-IA approach includes 10 impact categories, the ReCiPe 2016 Midpoint method includes 18, and the TRACI 2.1 method includes ten. Reporting conventions for impact categories vary, so the closest match within the TRACI 2.1 methodology was used in Table 3. For instance, "life cycle carbon emissions" were generalized as GWP.

Some studies focus on a single environmental indicator such as GWP (42%), for example, Llatas et al. (2022), Gao et al. (2024), Lu and Deng (2023); while others cover multiple environmental impact indicators (58%) (Zhou, Tam, et al., 2023a; Zhou, Tam, et al., 2023b; Bueno and Fabricio, 2018; (Abbasi and Noorzai, 2021), such as acidification potential (AP), eutrophication potential (EP), ozone depletion potential (ODP), smog formation potential (SFP), photochemical oxidant creation potential (POCP), abiotic depletion potential—elements (ADPE), abiotic depletion potential—fossil fuels (ADPF), primary energy demand (PED), primary non-renewable energy demand (PEDnr), and primary renewable energy demand (PEDr).

The *EeBGuide Handbook* delineates three tiers of comprehensive Building LCA: (1) Screening LCA, which constitutes a basic estimation involving simplified input and output data considering a few environmental impact indicators; (2) Simplified LCA, which employs specialised software tools such as Athena Impact Estimator; and (3) Complete LCA, an advanced analysis using advanced LCA software such as Simapro (Soust-Verdaguer et al., 2017). In this review, 12 articles prepared a Screening LCA, and 52 articles a Simplified LCA (Table 3).

EN 15978:2011 categorises four building life-cycle stages: product and construction (A1–A5), use (B1–B5), end-of-life (C1–C4), and benefits and loads beyond the system boundary (D). The selected stages are still highly related to the goal and scope of the study. Most articles focused on the production and construction phase (A, 92%). Other phases included were the use (B, 72%),the end-of-life (C, 70%), and benefits and loads beyond the system boundary (D, 40%) (Table 3).

Most references use common LCA databases, such as GaBi (Zhou Tam, et al., 2023a; Ma et al., 2024), Athena; the Inventory of Carbon and Energy (ICE) (Sharif and Hammad, 2019); Ecoinvent (Heydari and Heravi 2023), while others adopted a local database, such as the Ökobaudat (Forth, Hollberg, et al., 2023), and (Forth,Abualdenien, et al. 2023), or the Chinese Life Cycle Inventory Database (Zhang et al., 2022).

Furthermore, 40% of the publications do not consider indicators in addition to those set out in the LCA, such as design for disassembly (DfD) and adaptability (DfA), life-cycle cost (LCC), social cost, or waste production and management. Examples of studies that take these additional indicators into account include Lu et al. (2023), who proposed a method based on OpenBIM to calculate the social

Table 3 Sixty-five publications categorized according to LCA calculation.

References	*LCA Calculation Standard*	*LCA Database*	*LCA Application*	*Environmental Impact Indicators*	*A*	*B*	*C*	*D*	*Other Indicators*
(Zhou, Tam, et al., 2023a)	ISO 14040, ISO 14044	GaBi 2018 database	Simplified	TRACI 2.1	[A1-A3]	[B2–B6]	[C2–C4]	✓	–
(Zhou, Tam, et al., 2023b)	ISO 14040, ISO 14044	GaBi 2018 database	Simplified	TRACI 2.1	[A1-A3]	[B2–B6]	[C2–C4]	✓	–
(Bueno and Fabricio, 2018)	ISO 14040, ILCD Handbook	GaBi 2018 database	Simplified	TRACI 2.1	[A1-A3]	[B2–B6]	[C2–C4]		–
(Sharif and Hammad, 2019)	Not specified	Bath ICE and Athena database	Simplified	GWP	NS	NS	NS	NS	LCC
(Taher and Elbeltagi, 2023)	ISO 14040	Ecoinvent database	Simplified	Eco point method	[A1-A4]	[B2–B6]	[C2–C4]	–	–
(Llatas et al., 2022)	ISO 14040	Not specified	Screening	GWP	–	–	[C3–C4]	–	–
(Abbasi and Noorzai, 2021)	EN 15978	Athena database	Simplified	TRACI 2.1	[A1-A4]	[B2, B4, B6]	✓	✓	–
(Abdelaal et al., 2023)	EN 15978	OneClick LCA database	Screening	GWP, ODP, AP, EP, POCP	[A1-A4]	[B2, B4, B6]	✓	✓	LCC
(Sobhkhiz et al., 2021)	ISO 14040, ISO 14044	Not specified	Screening	GWP	[A1-A4]	–	–	–	LCC
(Abouhamad and Abu-Hamd, 2020)	ISO 14040 and EN 15978.	Athena database	Simplified	US EPA TRACI	[A1-A4]	[B2, B4, B6]	✓	✓	–

Contd.

Table 3 *Contd.*

References	*LCA Calculation Standard*	*LCA Database*	*LCA Application*	*Environmental Impact Indicators*	*A*	*B*	*C*	*D*	*Other Indicators*
(Falegari and Shirzadi Javid, 2022)	ISO 14040 and EN 15978.	OneClick LCA database	Screening	GWP, ODP, AP, EP	[A1-A4]	[B6]	✓	✓	–
(Su et al., 2021)	ISO 14040, ISO 14044	CLCD (Chinese Life Cycle Database)	Screening	GWP, AP, EP, RD	NS	NS	NS	NS	–
(Tushar, Bhuiyan, et al., 2022)	ISO 14040, ISO 14044	GaBi 2018 database	Simplified	GWP, ODP, AP, EP, SFP, PED, NRE, RED	[A1-A4]	[B2, B4, B6]	✓	✓	LCC
(Lu and Deng, 2023)	ISO 21931-1	Ökobaudat Database	Simplified	GWP	[A1-A4]	[B2 _B4]	✓	✓	Social Cost
(Santos et al., 2020)	EN 15978	Not specified	Complete	CML 2001	[A1-A4]	[B2-B4]	✓	✓	LCC
(Cheng et al., 2022)	ISO 14040, EN 15978	Not specified	Simplified	GWP, FFD, MRD, AP, POCP, EP, FE	[A1-A4]	[B4]	✓	–	–
(Ajayi et al., 2015)	ISO 14040, EN 15978	Athena database	Simplified	GWP, HT	[A1-A4]	[B4]	✓	✓	–
(Ma et al., 2024)	ISO 14040, ISO 14044, ISO 21930, ISO 21931, EN 15804, EN 15978	GaBi 2018 database	Simplified	GWP, AP, EP, SF, ODP, PED, NRE, REU	[A1-A4]	[C1–C4]	✓	✓	–

Contd.

Table 3 *Contd.*

References	*LCA Calculation Standard*	*LCA Database*	*LCA Application*	*Environmental Impact Indicators*	*A*	*B*	*C*	*D*	*Other Indicators*
(Jafari et al., 2024)	ISO 14040, EN 15978	Ecoinvent database	Simplified	ReCiPe 2016	[A1-A5]	[B6]	✓	✓	–
(Gao et al., 2024)	ISO 14040	eFootprint database	Simplified	GWP	[A1-A5]	–	–	–	–
(Chen and Gallardo, 2023)	ISO 14040, EN 15978	GaBi 2018 database and EPDs	Simplified	TRACI 2.1	NS	NS	NS	NS	LCC
(Greer et al., 2024)	ISO 14040, EN 15978	GaBi 2018 database	Simplified	GWP, AP, EP, ODP, SFP, PED, PEDnr, PEDr	[A1-A4]	[B2-B5]	✓	✓	–
(Forastiere et al., 2023)	EN 15978	GaBi and Ecoinvent databases	Simplified	GWP	[A1-A4]	[B1-B3, B6, B7]	–	–	LCC, Social Cost of Carbon
(Basbagill et al., 2013)	EN 15978, GBT 51366-2019	Ecoinvent and Athena datadases	Simplified	GWP	[A1-A3]	[B2-B5, B6]	✓	✓	–
(LLatas et al., 2022)	ISO 21931-2, ISO 14040, EN 15978	Ecoinvent v3.7.1 and the Spanish BEDEC	Screening	GWP	[A1-A5]	–	[C1–C2, C4]	–	LCC, Social Impact
(Soust-Verdaguer, Gutiérrez Moreno, et al., 2023)	ISO 21931-2, ISO 14040, EN 15978	Ecoinvent v3.7.1 and the Spanish BEDEC	Screening	GWP	[A1-A3, A5]	–	[C1–C2, C4]	–	LCC, Social Impact

Contd.

Table 3 *Contd.*

References	*LCA Calculation Standard*	*LCA Database*	*LCA Application*	*Environmental Impact Indicators*	*A*	*B*	*C*	*D*	*Other Indicators*
(Soust-Verdaguer, Gutiérrez, et al., 2023)	ISO 21931-2, ISO 14040, EN 15978	Ecoinvent v3.7.1 and the Spanish BEDEC	Screening	GWP	[A1-A3, A5]	–	[C1–C2, C4]		LCC, Social Impact
(Ayman Mohamed et al., 2023)	EN 15643	ICE database and EPDs	Simplified	GWP	[A1-A3]	[B6]	–	–	–
(Heydari and Heravi, 2023)	Not mentioned	ICE database and EPDs	Simplified	GWP	[A1-A3]	[B6]	–	–	LCC, Social Cost of Carbon
(Fenz et al., 2023)	EN 15978	Ökobaudat database	Simplified	GWP, AP, ODP, ADPE, EP, ADPF, POCP	[A1-A4]	[B2-B5, B6]	✓	✓	LCC, Urban Planning
(Hunt and Osorio-Sandoval, 2023)	NS	ICE database	Screening	GWP	[A1-A3]	[B6]	–	–	–
(Serrano-Baena et al., 2023)	EN 15804, SO_15804	ECO-Platform (EPD), Construction Price Generator	Screening	GWP, EE	NS	NS	NS	NS	LCC, Waste Mass
(Shibata et al., 2023)	EN 15978 and RICS	OneClick LCA database	Screening	GWP	[A1-A4]	[B1-B4, B6]	–	–	LCC

Contd.

Table 3 *Contd.*

References	*LCA Calculation Standard*	*LCA Database*	*LCA Application*	*Environmental Impact Indicators*	*A*	*B*	*C*	*D*	*Other Indicators*
(Boje et al., 2023)	EN 15978	Ecoinvent database, PSILCA 1.0 database	Simplified	EF v3.0	[A1-A3]	[B6, B7]	–	–	LCC, S-LCA
(Arvizu-Piña et al., 2023)	EN 15978	Mexicaniuh and EPDs	Screening	GWP, ADP, ODP, AP, WSP, POCP	[A1-A4]	[B4, B6]	[C1]	–	–
(Forth, Hollberg, et al., 2023)	Not mentioned	Ökobaudat database	Simplified	GWP	[A1-A3]	[B4] [C3	[C3, C4]	✓	–
(Forth, Abualdenien, et al., 2023)	ISO 14040, EN 15643–2	Ökobaudat database	Simplified	GWP	[A1-A3]	[B4] [C3	[C3, C4]	✓	–
(Schneider-Marin et al., 2022)	ISO 14040, EN 15804, EN 15978	Ökobaudat database	Simplified	GWP	[A1-A3]	[C3, C4]	[C3, C4]	✓	–
(Mowafy et al., 2023)	ISO 14040, EN 15804, EN 15978	KBOB database	Simplified	GWP	[A1-A3]	[B4, B6]	[C3, C4]	–	–
(Newberry et al., 2023)	ISO 14040 and EN 15978.	OneClick LCA database	Simplified	GWP	[A1-A5]	[B4, B6,	✓	–	–
(Atashbar and Noorzai, 2023)	ISO 14040 and EN 15978.	Athena database	Simplified	GWP, PED	A1-A5]	[B2, B4]	✓	✓	–
(Deng and Lu, 2023)	ISO 14040 and EN 15978.	CEF database	Simplified	GWP	[A1-A5]	[B1- B5]	✓	✓	–

Contd.

Table 3 *Contd.*

References	*LCA Calculation Standard*	*LCA Database*	*LCA Application*	*Environmental Impact Indicators*	*A*	*B*	*C*	*D*	*Other Indicators*
(Hansen et al., 2023; Kim and Kim, 2023)	EN 15978	Ökobaudat database	Simplified	GWP	[A1-A3]	[B4]	[C3, C4]		–
(Kim and Kim, 2023)	EN 15804	EPDs, KICT	Simplified	GWP	[A1-A5]	–	✓	✓	Cost, DfD,
(Carvalho et al., 2023)	Not specified	Not specified	Simplified	GWP, ODP, AP, POCP, EP, FFDP	[A1-A3]	[B6]	[C3, C4]	✓	Cost
(Morsi et al., 2022)	EN 15978, EN 15804, ISO 21931-1, ISO 21929-1, and ISO 14040	OneClick LCA database	Simplified	GWP, AP, EP, ODP, POCP, PED	[A1-A5]	[B1-B7]	✓	✓	Time, Cost
(Nehasilová et al., 2022)	EN 15978, EN 15804, ISO 21931-1, ISO 21929-1, and ISO 14040	Ecoinvent database	Simplified	GWP, AP, EP, ODP, POCP, PED	[A1-A3]	–	–	–	Cost
(Tushar, Zhang, et al., 2022)	ISO 14040, ISO 14044	Ecoinvent 3.8 database	Simplified	GWP, AP, EP, ODP, SFP, PED, NRED, RED	[A1-A3]	–	–	–	–
(Zhang et al., 2022)	ISO 14040, ISO 14044	CLCD (Chinese Life Cycle Database)	Simplified	GWP	[A1-A5]	[B6]	–	–	Cost, CEI

Contd.

Table 3 *Contd.*

References	*LCA Calculation Standard*	*LCA Database*	*LCA Application*	*Environmental Impact Indicators*	*A*	*B*	*C*	*D*	*Other Indicators*
(Besana and Tirelli, 2022)	EN 15978	OneClick LCA database	Simplified	GWP	[A1-A5]	–	–	–	–
(Motalebi et al., 2022)	EN 15978	Ecoinvent 3.8 database	Simplified	GWP, AP, EP, SFP, NRE	[A1-A5]	[B2-B5, B6]	[C2-C4]	–	LCC
(Xu et al., 2022)	ISO 14044, PAS 2050, EN 15978	SimaPro databases	Simplified	GWP	[A1-A5]	–	–	–	–
(Dauletbek and Zhou, 2022)	EN 15978	Ecoinvent 3.8 database	Simplified	GWP, AP, EP, SFP, NRE	[A1-A5]	[B2-B5, B6]	[C2-C4]	–	LCC
(Najjar et al., 2022)	ISO 14040, ISO 14044	Ecoinvent database	Simplified	IMPACT 2002+ and ILCD 2011	[A1-A5]	[B2-B5]	[C2-C4]	–	–
(Alwan and Ilhan Jones, 2022)	EN 15978	ICE database	Simplified	GWP	[A1-A3]	–	–	–	–
(Filho et al., 2022)	EN 15978	GaBi 2018 database	Simplified	GWP, AP, EP, ODP, POCP, SFP, PED	[A1-A4]	[B2-5]	[C2-C4]	✓	Cost, Social Impact
(Soust-Verdaguer et al., 2022)	EN 15978	BEDEC database	Simplified	GWP	[A1-A3, A5]	–	–	–	Cost, Social Impact
(Martínez-Rocamora et al., 2021)	EN 15978	GaBi 2018 database	Simplified	GWP, AP, EP, ODP, SFP, PED, NRED, RED	[A1-A4]	[B2, B4]	[C2-C4]	✓	–

Contd.

Table 3 *Contd.*

References	*LCA Calculation Standard*	*LCA Database*	*LCA Application*	*Environmental Impact Indicators*	*A*	*B*	*C*	*D*	*Other Indicators*
(Figueiredo et al., 2021)	ISO 14040, ISO 14044	GaBi 2018 database	Simplified	GWP, AP, EP	[A5]	[B2 - B5, B6]	✓	–	LCC, Fair Wage Potential
(Ansah et al., 2021)	ISO 14040, ISO 14044	Ecoinvent database	Simplified	GWP, AP, EP, ODP, POCP, HTP, MAEP, TETP, RE	[A1-A5]	[B2, B4]	✓	–	–
(Alwan et al., 2021)	ISO 14040, EN 15978	ICE database	Simplified	GWP	[A1-A3]	–	–	–	–
(Lee et al., 2021)	ISO 14040, ISO 14044	EPDs	Simplified	GWP, AP, EP, ODP, ADP, PDCP	[A1-A5]	[B2, B4]	✓	–	–
(Uddin et al., 2021)	Not specified	ICE database	Simplified	GWP	[A1-A3]	[B2, B4]	✓	–	–
(Carvalho et al., 2021)	EN 15978	Ecoinvent database	Simplified	GWP, AP, EP, ODP, ADP, POCP, ENR	[A1-A3]	[B2, B4]	✓	–	LCC
(Tushar et al., 2021a)	Not specified	GaBi 2018 database	Simplified	GWP, AP, EP, ODP, SFP, PED, NRED, RED	[A1-A3]	[B2, B4]	✓	–	–

cost of carbon through the building life cycle; Santos et al. (2020), who developed a method to integrate BIM with LCA and LCC, incorporating environmental, economic, and physical information into BIM objects; Llattas et.al. (2022), who proposed a method to identify the design solution with the lowest CO2 emissions, the lowest cost, and the most beneficial impact on local employment; Fenz et al. (2023), who developed a web-based tool to assess the performance of renovation scenarios in terms of energy efficiency, LCA, LCC, and urban planning key performance indicators related to social well-being; Serrano-Baena et al. (2023), who developed a BIM-based software designed to assess the total economic cost, GWP, embodied energy, and waste mass; and, Kim et al. (2023), who proposed an LCA-based LCA-DfD method to quantitatively calculate environmental loads based on various design and deconstruction scenarios.

There is, therefore, a lack of tools to assess how circular economy principles such as DfD, waste management, and DfA influence LCA results. DfD and waste management enable the reuse of components, products, and materials, affecting the end-of-life (C) and benefits and loads beyond the system boundary (D) stages, and DfA extends the useful lifespan of the building by easily adapting it to future functional and user requirements.

3.3 BIM-LCA Integration

According to Zheng et al. (2023), the classification of BIM-LCA integration can be broadly categorised according to three principles: data exchange, data extraction, and data flow. Safari and AzariJafari (2021) defined three categories for data exchange processes between BIM and LCA tools: conventional, static, and dynamic. In the conventional method, the bill of quantities or materials is extracted from the BIM model, typically as an Excel or CSV file. The collected data is then entered manually into the LCA software. In the static approach, the BIM model is converted into an IFC file and imported into the LCA software. This method utilizes the information delivery specifications (IDS) of BIM objects (BuildingSMART International, n.d.) for data sharing, which simplifies model updates and eliminates the need to recombine environmental information for each material or product. The dynamic approach allows calculations to be performed directly within the BIM environment using add-ons, facilitating an iterative design process.

Soust-Verdaguer et al. (2017) and Wastiels and Decuypere (2019) identified five distinct types of data flow: (a) export the BoQ from the BIM tool to LCA tools, as followed by Simba Pro, eFootprint LCA, OneClick LCA, Athena Impact Estimator, Athena Eco calculator, LCAbyg, OpenLCA (Table 4 (b)) export BIM models in IFC format and then load them into the LCA tool and align them with an established LCA data; (c) use a BIM viewer to handle data and then transfer it to an LCA software; (d) employ plug-ins that enable LCA analysis within the BIM application, such as the Autodesk Tally, Bombyx, Revit Dynamo scripts; (e) embed LCA information directly into the BIM objects, as suggested by Santos et al. (2020).

Table 4 Characterisation of the LCA software used in the 65 publications.

LCA Software	*Type of Software*	*Regionality*	*Database*	*New Data in the Database*	*BoM or BoQ Import*	*Design Comparisons*	*References*
Bombyx	Grasshopper add-on	KBOB database	Swiss LCA database	Yes	Add-on to pull the Revit element to Grasshopper	Yes. Material options	(Mowafy et al., 2023)
Athena Impact Estimator	Third-party dedicated LCA tool	North America	Athena database and EPDs	No	BoM from CSV, XML, Excel ... Allows manual entries without a model.	Yes. Design options don't need to be modelled.	(Abbasi and Noorzai, 2021; Abouhamad and Abu-Hamd, 2020; Ajayi et al., 2015; Atashbar and Noorzai, 2023)
Athena Eco calculator	Third-party dedicated LCA tool	North America	Athena database and EPDs	No	BoM from CSV, XML, Excel ... Allows manual entries without a model.	Yes. Design options don't need to be modelled.	(Basbagill et al., 2013)
Autodesk Tally	Autodesk Revit add-on	North America	GaBi 2018 database and EPDs	No	BOM from Revit API. No manual entries without a model	Yes, but needs Revit design options.	(Bueno and Fabricio, 2018; Chen and Gallardo, 2023; Figueiredo et al., 2021; Filho et al., 2022; Greer et al., 2024; Ma et al., 2024; Martínez-Rocamora et al., 2021; Motalebi et al., 2022; Tushar, Bhuiyan, et al., 2022; Tushar et al., 2021a; Tushar, Zhang, et al., 2022; Zhou, Tam, et al., 2023a, 2023b)

Contd.

Table 4 *Contd.*

LCA Software	*Type of Software*	*Regionality*	*Database*	*New Data in the Database*	*BoM or BoQ Import*	*Design Comparisons*	*References*
eFootprint LCA	Third-party dedicated LCA tool	China	CLCD-China	No	BoM from CSV, XML, Excel ... Allows manual entries without a model.	Yes. Design options don't need to be modelled	(Gao et al., 2024)
LCAbyg	Third-party dedicated LCA tool	Denmark	Ökobaudat database	No	BoM from CSV, XML and JSON files.	Yes. Design options don't need to be modelled	(Hansen et al., 2023)
OneClick LCA	Third-party dedicated LCA tool and an add-on to export data from Revit	North and South America, Europe, Middle East, Asia	Multiple databases	No	Revit, IES-VE, Design Builder, Excel, CSV, gbXML, SketchUp Pro, ArchiCAD, and more.	Yes. Design options don't need to be modelled.	(Abdelaal et al., 2023; Besana and Tirelli, 2022; Falegari and Shirzadi Javid, 2022; Forastiere et al., 2023; Morsi et al., 2022; Newberry et al., 2023)
Open LCA	Third-party dedicated LCA tool	Germany	Multiple databases	No	Excel, CSV	Yes. Design options don't need to be modelled.	(Najjar et al., 2022)
Python script Brightway2 libraries	Python script	Luxembourg	Ecoinvent database, PSILCA 1.0 database	No	IFC	Yes, but needs Revit design options.	(Boje et al., 2023)

Contd.

Table 4 *Contd.*

LCA Software	*Type of Software*	*Regionality*	*Database*	*New Data in the Database*	*BoM or BoQ Import*	*Design Comparisons*	*References*
Sima Pro	Third-party dedicated LCA tool	Netherlands	Multiple databases	Yes	Excel, CSV	Yes. Design options don't need to be modelled.	(Basbagill et al., 2013; Jafari et al., 2024; Taher and Elbeltagi, 2023; Xu et al., 2022)
eBalance	Third-party dedicated LCA tool	China	Multiple databases	No	Excel, CSV	Yes. Design options don't need to be modelled.	(Su et al., 2021)
Tool from (LLatas et al., 2022)	Dynamo script	Spain	Ecoinvent database	No	Revit	Yes	(Soust-Verdaguer, Gutiérrez, et al., 2023; Soust-Verdaguer, Gutiérrez Moreno, et al., 2023)
WEBlca	Third-party dedicated LCA tool	China	Chinese Life Cycle Inventory Database	No	Excel, CSV	Yes. Design options don't need to be modelled.	(Dauletbek and Zhou, 2022)
Tools developed the authors	varies	varies	varies	varies	varies	varies	-

Contd.

The LCA tools were categorized into the five categories mentioned above, and it was found that 60% of the references studied are third-party LCA tools, 23% are add-ons to BIM software, mostly Revit, 9% involve parametric modelling, such as Dynamo or Grasshopper scripts, and 8% feature LCA embedded in BIM objects, where the environmental information is parameterized in the BIM objects before calculation.

About data exchange, we found that 40% of the tools use conventional data exchange methods, 38% use dynamic data exchange, and 22% use static data exchange (Figure 5). Additionally, in 89% of the proposed methods, the correspondence between environmental data and the BoQ is carried out manually. Only 5% use an Excel sheet with the Family and Type names corresponding to the LCA values (Alwan et al., 2021; Ansah et al., 2021; Carvalho et al., 2021), and 3% use an automatic learning algorithm, such as Natural Language Processing, to establish this correspondence automatically with good precision (Forth, Abualdenien, et al., 2023; Forth, Hollberg, et al., 2023), and 2% use the construction classification system to link the elements classified in the BoQ (Parece, Resende, et al., 2024) (Table 4).

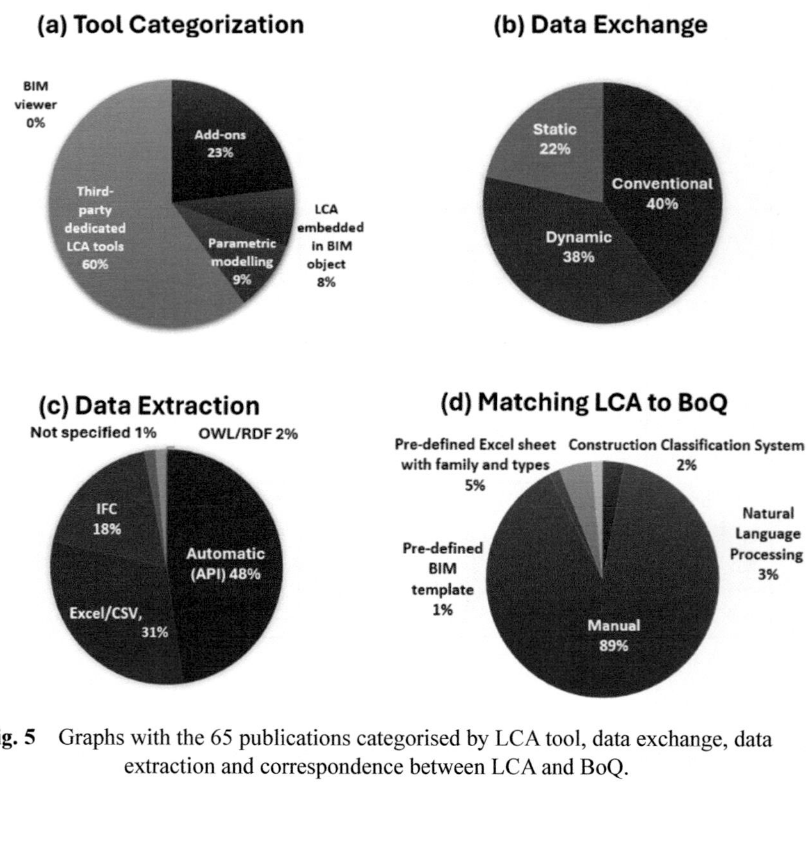

Fig. 5 Graphs with the 65 publications categorised by LCA tool, data exchange, data extraction and correspondence between LCA and BoQ.

3.4 Case Study Application

Most publications selected an entire building as the functional unit for developing the LCA, although some only considered part of it (Tushar, Bhuiyan, et al., 2022, Bueno and Fabricio, 2018; Jafari et al., 2024). Concerning building properties, 37% focused on multi-residential buildings, 18% on single-dwellings and 9% on office buildings. Furthermore, 85% of the articles focused on new-build projects, while only 14% focused on renovation projects (Figure 6).

The results indicated that buildings assessed in the articles were distributed across different countries and regions, with 18% located in mainland China.

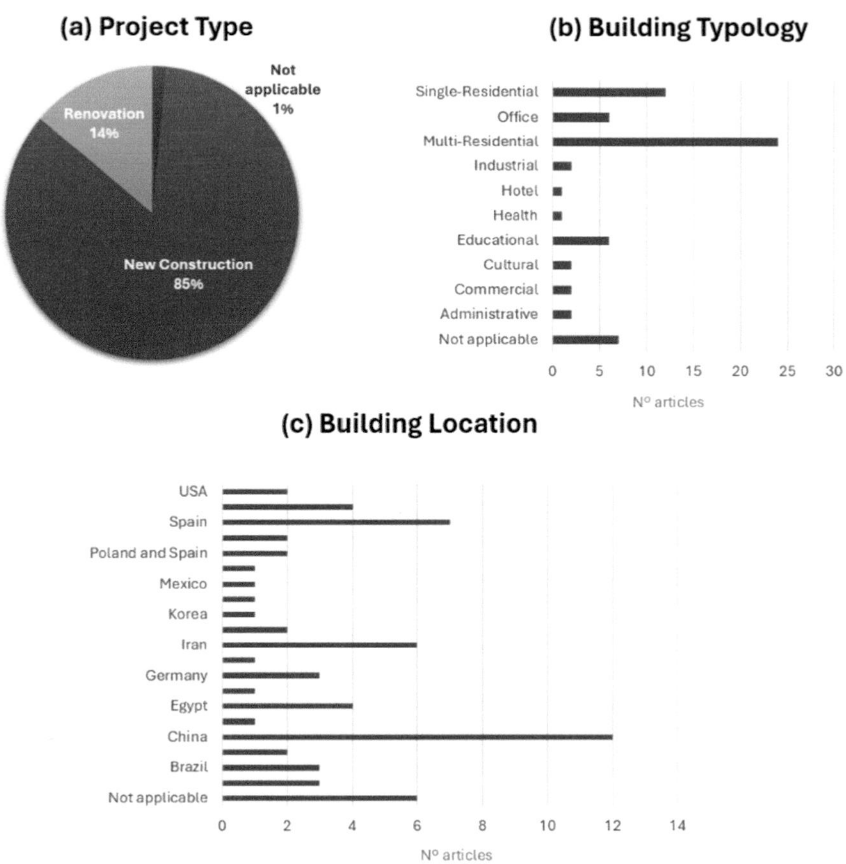

Fig. 6 Charts with the 65 publications categorised by Project Type, Building Typology, and Building Location.

3.5 Decision Support

Decisions made during a project significantly impact a building's performance over its entire lifespan. While project goals may conflict in the short and medium term, they create complex challenges. Optimisation entails identifying a function's highest or lowest value while adhering to specified constraints by adjusting multiple variables. In building design, multi-objective optimisation is often employed to tackle conflicting objectives that impede the simultaneous optimisation of all goals (Hassan et al., 2022). Various multi-objective optimisation methods, such as evolutionary algorithms, particle swarm optimisation, and the Non-Dominated Sorting Genetic Algorithm (NSGA), are applicable in this context (Hassan et al., 2024). For example, Genetic algorithms are used for analysing the trade-offs between economic and environmental factors (Martínez-Rocamora et al., 2021; Mowafy et al., 2023) refurbishment and renovation optimisation (Motalebi et al., 2022; Sharif and Hammad, 2019), for the tradeoff between embodied and operational energy (Abbasi and Noorzai, 2021), and building envelope optimisation (Atashbar and Noorzai, 2023; Tushar, Zhang, et al., 2022), for predicting the environmental impact of different building typologies (Martínez-Rocamora et al., 2021).

Multi-Criteria Decision Analysis (MCDA) is another decision-support method. Various multi-criteria methods have been developed and applied in construction design. For example, Domingos and Rato (2019) used the MACBETH tool to find the best-fit construction materials for extreme climate scenarios, while Stellacci et al. (2024) used parametric modelling tools and multi-criteria decision analysis to find the best retrofit scenario considering energy consumption and architectural configuration. Taher et al. (2023) developed a decision support framework using the analytical hierarchy process (AHP) to choose the appropriate design alternative for sustainable buildings considering cost, time, aesthetics, availability of material and energy efficiency. Abdelaal et al. (2023) aid in ranking and selecting the type of concrete based on sustainability criteria including CO2 emissions, embodied energy, and cost using the AHP method. Filho et al. (2022) used the Fuzzy AHP to find the best design option considering environmental and socio-political criteria.

Moreover, sensitivity analysis, regression techniques, and other probabilistic methods are used to manage uncertainties through sampling and have proven particularly effective in early LCA studies (Hassan et al., 2022). These methods allow users to explore various alternatives by iteratively evaluating design variables. Sensitivity analysis has been employed in early LCA studies to help reduce the embodied environmental impacts of building component materials and dimensioning choices (Basbagill et al., 2013; Gao et al., 2024; Heydari and Heravi, 2023; Tushar et al., 2021b) and to identify the influential factors in building LCA (Zhou, Tam, et al., 2023b) and analyse the importance of independent variables.

Despite the benefits for design-making of using the aforementioned methods, only 15% of the studies utilised the target optimisation methods, 7% employed MCDA, and 13% incorporated sensitivity analysis.

Conclusion

This study examined 65 articles on BIM-based LCA published between 2013 and March 2024. A systematic review following Prisma Guidelines was employed to answer the research question: What are the limitations, trends, and future developments of BIM-based LCA for building design?

The BIM-based LCA topic has received considerable attention during the last five years. The five hottest topics in the knowledge area have been identified as LCA, BIM, carbon emissions, global warming potential, and sustainability. Chinese researchers have been the key in the development of BIM and LCA in the construction sector, making significant contributions and serving as important players in the collaboration network.

Autodesk Revit and Graphisoft ArchiCAD are the most commonly used BIM software. Most studies do not specify the models LoD, though some identify LoD 300 as suitable for accurate LCC and LCA calculations, while others perform simplified LCA studies with LoD 200 BIM models.

Sixty per cent of the references studied are third-party LCA tools, 23% are add-ons to BIM software, mostly Revit, and 9% where environmental information is parameterised in BIM objects. Concerning data exchange, 40% of the tools use a conventional method. This procedure is inefficient because any changes require restarting the process from the beginning, making it prone to errors and time-consuming. Additionally, 89% of the proposed methods carry out the correspondence between environmental data and the BoQ manually.

Most of the references studied performed an LCA on new-build projects, and only 35% applied a decision support method such as multi-objective optimisation, MCDA, or sensitivity analysis.

Although we have identified many different approaches to integrating BIM and LCA, this process needs further development. Our findings reveal that the major literature gaps are:

(1) BIM-based LCA still requires many manual processes, such as editing the BoQ and mapping BIM objects or materials to environmental data. Both processes can be automated. The first can be addressed by adopting standardised modelling and information management strategies and using static or dynamic data exchange tools. The second can be automated by adopting a universal nomenclature, such as construction classification systems, to establish an automatic link between BIM objects and LCA databases, enabling the application of algorithms like Natural Language Processing

(2) There is a notable lack of tools that enable continuous and accurate monitoring of environmental impacts throughout all project phases. The available tools are typically adapted for either highly detailed design or early design. Future tools should align LCA databases with design phases to improve accuracy

and reliability, as suggested by Cavalliere et al. (2019) and Parece, Costa, et al. (2024)

(3) There is a notable lack of studies on how circular economy strategies, such as DfD and waste management, and DfA, influence the LCA results in building projects. DfD and waste management influence end-of-life scenarios, and DfA the use stage. There is a need to investigate how these stratagies are quantitatively reflected in the carbon emissions throughout the building's life cycle.

(4) There is a notable lack of methodologies adapted for renovation or refurbishment projects. These BIM models usually have different attributes, including unique modelling techniques, differentiation between existing and new structures and rehabilitation procedures, and most of the available LCA tools do not allow for interoperability and workflow automation in these cases.

(5) It is evident that there remains a dearth of research on decision support methods applied to LCA in building projects, considering various indicators that frequently conflict, such as cost, operational carbon, and embodied carbon. These decision support methods can include multi-objective optimisation or MCDA.

Acknowledgement

This research was supported by the doctoral Grant PRT/BD/154261/2022, financed by the Portuguese Foundation for Science and Technology (FCT), and with funds under the MIT Portugal Program, and partially founded by ISTAR projects: UIDB/04466/2020 and UIDP/04466/2020.

References

Abbasi, S. and Noorzai, E. (2021). The BIM-Based multi-optimisation approach in order to determine the trade-off between embodied and operation energy focused on renewable energy use. *Journal of Cleaner Production*, *281*, 125359. https://doi.org/10.1016/J.JCLEPRO.2020.125359.

Abdelaal, M.A., Seif, S.M., El-Tafesh, M.M., Bahnas, N., Elserafy, M.M. and Bakhoum, E.S. (2023). Sustainable assessment of concrete structures using BIM–LCA–AHP integrated approach. *Environment, Development and Sustainability*, 1–20. https://doi.org/10.1007/S10668-023-03701-3/FIGURES/7.

Abouhamad, M. and Abu-Hamd, M. (2020). Life Cycle Environmental Assessment of Light Steel Framed Buildings with Cement-Based Walls and Floors. *Sustainability*, *12*(24), 10686. https://doi.org/10.3390/SU122410686.

AIA (American Institute of Architects). (2013). *E203-2013 Building Information Modelling and Digital Data Exhibit*. AIA Washington, DC.

Ajayi, S.O., Oyedele, L.O., Ceranic, B., Gallanagh, M. and Kadiri, K.O. (2015). Life cycle environmental performance of material specification: A BIM-enhanced comparative assessment. *International Journal of Sustainable Building Technology and Urban Development*, *6*(1), 14–24. https://doi.org/10.1080/2093761X.2015.1006708.

Alwan, Z. and Ilhan Jones, B. (2022). IFC-based embodied carbon benchmarking for early design analysis. *Automation in Construction*, *142*, 104505. https://doi.org/10.1016/j.autcon.2022.104505.

Alwan, Z., Nawarathna, A., Ayman, R., Zhu, M. and ElGhazi, Y. (2021). Framework for parametric assessment of operational and embodied energy impacts utilising BIM. *Journal of Building Engineering*, *42*, 102768. https://doi.org/10.1016/j.jobe.2021.102768.

Ansah, M.K., Chen, X., Yang, H., Lu, L. and Lam, P.T.I. (2021). Developing an automated BIM-based life-cycle assessment approach for modularly designed high-rise buildings. *Environmental Impact Assessment Review*, *90*, 106618. https://doi.org/10.1016/j.eiar.2021.106618.

Arvizu-Piña, V.A., Armendáriz López, J.F., García González, A.A. and Barrera Alarcón, I.G. (2023). An open access online tool for LCA in building's early design stage in the Latin American context: A screening LCA case study for a bioclimatic building. *Energy and Buildings*, *295*, 113269. https://doi.org/10.1016/j.enbuild.2023.113269.

Atashbar, H. and Noorzai, E. (2023). Optimization of Exterior Wall Cladding Materials for Residential Buildings Using the Non-dominated Sorting Genetic Algorithm II (NSGAII) based on the Integration of Building Information Modelling (BIM) and Life Cycle Assessment (LCA) for Energy Consumption: A Case Study. *Sustainability*, *15*(21), 15647. https://doi.org/10.3390/su152115647.

Ayman Mohamed, R., Alwan, Z., Salem, M. and McIntyre, L. (2023). Automation of embodied carbon calculation in digital built environment tool utilizing UK LCI database. *Energy and Buildings*, *298*, 113528. https://doi.org/10.1016/j.enbuild.2023.113528.

Basbagill, J., Flager, F., Lepech, M. and Fischer, M. (2013). Application of life-cycle assessment to early stage building design for reduced embodied environmental impacts. *Building and Environment*, *60*, 81–92. https://doi.org/10.1016/j.buildenv.2012.11.009.

Besana, D. and Tirelli, D. (2022). Reuse and Retrofitting Strategies for a Net Zero Carbon Building in Milan: An Analytic Evaluation. *Sustainability*, *14*(23), 16115. https://doi.org/10.3390/su142316115.

Boje, C., Hahn Menacho, Á.J., Marvuglia, A., Benetto, E., Kubicki, S., Schaubroeck, T. and Navarrete Gutiérrez, T. (2023). A framework using BIM and digital twins in facilitating LCSA for buildings. *Journal of Building Engineering*, *76*, 107232. https://doi.org/10.1016/j.jobe.2023.107232.

Bueno, C. and Fabricio, M.M. (2018). Comparative analysis between a complete LCA study and results from a BIM-LCA plug-in. *Automation in Construction*, *90*, 188–200. https://doi.org/10.1016/J.AUTCON.2018.02.028.

Bueno, C., Pereira, L.M. and Fabricio, M.M. (2018). Life cycle assessment and environmental-based choices at the early design stages: An application using building information modeling. 1–15*Arch. Engg. And Design Mgt.*,*14*(40), 1–15. https://doi.org/10.1080/17452007.2018.1458593.

Carvalho, J.P., Bragança, L. and Mateus, R. (2023). Automating building sustainability assessment using building information modelling: A case study. *Journal of Building Engineering*, *76*, 107228. https://doi.org/10.1016/j.jobe.2023.107228.

Carvalho, J.P., Villaschi, F.S. and Bragança, L. (2021). Assessing Life Cycle Environmental and Economic Impacts of Building Construction Solutions with BIM. *Sustainability*, *13*(16), 8914. https://doi.org/10.3390/su13168914.

Cavalliere, C., Habert, G., Dell'Osso, G.R. and Hollberg, A. (2019). Continuous BIM-based assessment of embodied environmental impacts throughout the design process. *Journal of Cleaner Production*, *211*, 941–52. https://doi.org/10.1016/J.JCLEPRO.2018.11.247.

Chen, Y. and Gallardo, S. (2023). A Multi-Objective Optimization Method for the Design of a Sustainable House in Ecuador by Assessing LCC and LCEI. *Sustainability, 2024*, *16*(1), 168. https://doi.org/10.3390/SU16010168.

Cheng, B., Lu, K., Li, J., Chen, H., Luo, X. and Shafique, M. (2022). Comprehensive assessment of embodied environmental impacts of buildings using normalised environmental impact factors. *Journal of Cleaner Production*, *334*, 130083. https://doi.org/10.1016/J.JCLEPRO.2021.130083.

Cheung, F.K.T., Rihan, J., Tah, J., Duce, D. and Kurul, E. (2012). Early stage multi-level cost estimation for schematic BIM models. *Automation in Construction*, *27*, 67–77. https://doi.org/10.1016/J.AUTCON.2012.05.008.

Dauletbek, A. and Zhou, P. (2022). BIM-based LCA as a comprehensive method for the refurbishment of existing dwellings considering environmental compatibility, energy efficiency, and profitability:

A case study in China. *Journal of Building Engineering*, *46*, 103852. https://doi.org/10.1016/j.jobe.2021.103852.

Deng, X. and Lu, K. (2023). Multi-level assessment for embodied carbon of buildings using multi-source industry foundation classes. *Journal of Building Engineering*, *72*, 106705. https://doi.org/10.1016/j.jobe.2023.106705.

Domingos, L. and Rato, V. (2019). Multi-criteria material selection for buildings in challenging environments. *IOP Conference Series: Earth and Environmental Science*, *297*(1), 012038. https://doi.org/10.1088/1755-1315/297/1/012038.

Falegari, S. and Shirzadi Javid, A.A. (2022). Integrating building information modelling and life cycle assessment to analyse the role of climate and passive design parameters in energy consumption. *Energy and Environment*, *35*(4). https://doi.org/10.1177/0958305X221145923/ASSET/IMAGES/LARGE/10.1177_0958305X221145923-FIG10.JPEG.

Fenz, S., Giannakis, G., Bergmayr, J. and Iousef, S. (2023). RenoDSS: A BIM-based building renovation decision support system. *Energy and Buildings*, *288*, 112999. https://doi.org/10.1016/j.enbuild.2023.112999.

Figueiredo, K., Pierott, R., Hammad, A.W.A. and Haddad, A. (2021). Sustainable material choice for construction projects: A Life Cycle Sustainability Assessment framework based on BIM and Fuzzy-AHP. *Building and Environment*, *196*, 107805. https://doi.org/10.1016/j.buildenv.2021.107805.

Filho, M.V.A.P. M., da Costa, B.B.F., Najjar, M., Figueiredo, K.V., de Mendonça, M.B. and Haddad, A.N. (2022). Sustainability Assessment of a Low-income Building: A BIM-LCSA-FAHP-based Analysis. *Buildings*, *12*(2), 181. https://doi.org/10.3390/buildings12020181.

Forastiere, S., Piselli, C., Pioppi, B., Balocco, C., Sciurpi, F. and Pisello, A.L. (2023). Towards Achieving Zero Carbon Targets in Building Retrofits: A Multi-Parameter Building Information Modeling (BIM) Approach Applied to a Case Study of a Thermal Bath. *Energies*, *16*(12), 4757. https://doi.org/10.3390/en16124757.

Forth, K., Abualdenien, J. and Borrmann, A. (2023). Calculation of embodied GHG emissions in early building design stages using BIM and NLP-based semantic model healing. *Energy and Buildings*, *284*, 112837. https://doi.org/10.1016/j.enbuild.2023.112837.

Forth, K., Hollberg, A. and Borrmann, A. (2023). BIM4EarlyLCA: An interactive visualization approach for early design support based on uncertain LCA results using open BIM. *Developments in the Built Environment*, *16*, 100263. https://doi.org/10.1016/j.dibe.2023.100263.

Gao, Y., Wang, J. and Yiu, T.W. (2024). Multi-information integration-based life-cycle analysis of greenhouse gas emissions for prefabricated construction: A case study of Shenzhen. *Environmental Impact Assessment Review*, *104*, 107330. https://doi.org/10.1016/J.EIAR.2023.107330.

Greer, F., Raftery, P., Brager, G., -, al, Hernández-Fernández, J.D., Vílchez, J.M., Iglesias-Páramo, J., McCord, K.H., Dillon, H.E., Gunderson, P., Carlson, S., Phillips, A.R., Griechen, D. and Antonopoulos, C.A. (2024). Strategies for connecting whole-building LCA to the low-carbon design process. *Environmental Research: Infrastructure and Sustainability*, 4(1), 015002. https://doi.org/10.1088/2634-4505/AD17CEHansen, R.N., Hoxha, E., Rasmussen, F.N., Ryberg, M.W., Andersen, C.E. and Birgisdóttir, H. (2023). Enabling rapid prediction of quantities to accelerate LCA for decision support in the early building design. *Journal of Building Engineering*, *76*, 106974. https://doi.org/10.1016/j.jobe.2023.106974.

Hassan, S.R., Megahed, N.A., Abo Eleinen, O.M. and Hassan, A.M. (2022). Toward a national life cycle assessment tool: Generative design for early decision support. *Energy and Buildings*, *267*, 112144. https://doi.org/10.1016/J.ENBUILD.2022.112144.

Hassan, S.R., Megahed, N.A., Eleinen, O.M.A. and Hassan, A.M. (2024). An Overview of LCA Integration Methods at the Early Design Stage towards National Application. *In*: *Proceedings of IWBBIO Conference on Enginnering Solutions towards Sustainable Development*, 357–74. https://doi.org/10.1007/978-3-031-46491-1_22.

Heydari, M. and Heravi, G. (2023). A BIM-based framework for optimization and assessment of buildings' cost and carbon emissions. *Journal of Building Engineering*, *79*, 107762. https://doi.org/10.1016/j.jobe.2023.107762.

Hollberg, A., Genova, G. and Habert, G. (2020). Evaluation of BIM-based LCA results for building design. *Automation in Construction*, *109*, 102972. https://doi.org/10.1016/J.AUTCON.2019.102972.

Hunt, J. and Osorio-Sandoval, C.A. (2023). Assessing Embodied Carbon in Structural Models: A Building Information Modelling-based Approach. *Buildings*, *13*(7), 1679. https://doi.org/10.3390/buildings13071679.

Information Delivery Specifications (IDS). (n.d.). *buildingSMART International*. Retrieved May 24, 2024, from https://www.buildingsmart.org/standards/bsi-standards/information-delivery-specifications-ids/

ISO-14040, I.S.O. 14040. (2006). *Environmental Management–Life Cycle Assessment–Principles and Framework*.

Jafari, M., Khoshand, A., Sadeghi, N. and Mirzanagh, P.A. (2024). A comparative LCA of external wall assemblies in context of Iranian market: Considering embodied and operational energy through BIM application. *Environmental Science and Pollution Research International*, *31*(5), 7364–79. https://doi.org/10.1007/S11356-023-31451-2/FIGURES/8.

Kim, S. and Kim, S.-A. (2023). A design support tool based on building information modeling for design for deconstruction: A graph-based deconstructability assessment approach. *Journal of Cleaner Production*, *383*, 135343. https://doi.org/10.1016/j.jclepro.2022.135343.

Lee, S., Tae, S., Jang, H., Chae, C.U. and Bok, Y. (2021). Development of Building Information Modeling Template for Environmental Impact Assessment. *Sustainability*, *13*(6), 3092. https://doi.org/10.3390/su13063092.

Llatas, C., Quiñones, R. and Bizcocho, N. (2022). Environmental Impact Assessment of Construction Waste Recycling versus Disposal Scenarios using an LCA-BIM Tool during the Design Stage. *Recycling*, *7*(6), 82. https://doi.org/10.3390/RECYCLING7060082/S1.

LLatas, C., Soust-Verdaguer, B., Hollberg, A., Palumbo, E. and Quiñones, R. (2022). BIM-based LCSA application in early design stages using IFC. *Automation in Construction*, *138*, 104259. https://doi.org/10.1016/j.autcon.2022.104259.

Lu, K. and Deng, X. (2023). OpenBIM-based assessment for social cost of carbon through building life cycle. *Sustainable Cities and Society*, *99*. https://doi.org/10.1016/j.scs.2023.104871.

Ma, L., Azari, R. and Elnimeiri, M. (2024). A Building Information Modelling-based Life Cycle Assessment of the Embodied Carbon and Environmental Impacts of High-Rise Building Structures: A Case Study. *Sustainability (Switzerland)*, *16*(2). https://doi.org/10.3390/su16020569.

Martínez-Rocamora, A., Rivera-Gómez, C., Galán-Marín, C. and Marrero, M. (2021). Environmental benchmarking of building typologies through BIM-based combinatorial case studies. *Automation in Construction*, *132*, 103980. https //doi.org/10.1016/j.autcon.2021.103980.

Morsi, D.M.A., Ismaeel, W.S.E., Ehab, A. and Othman, A.A.E. (2022). BIM-based life-cycle assessment for different structural system scenarios of a residential building. *Ain Shams Engineering Journal*, *13*(6), 101802. https://doi.org/10.1016/j.asej.2022.101802.

Motalebi, M., Rashidi, A. and Nasiri, M.M. (2022). Optimization and BIM-based life-cycle assessment integration for energy efficiency retrofit of buildings. *Journal of Building Engineering*, *49*, 104022. https://doi.org/10.1016/j.jobe.2022.104022.

Mowafy, N., El Zayat, M. and Marzouk, M. (2023). Parametric BIM-based life-cycle assessment framework for optimal sustainable design. *Journal of Building Engineering*, *75*, 106898. https://doi.org/10.1016/j.jobe.2023.106898.

Najjar, M.K., Figueiredo, K., Evangelista, A.C.J., Hammad, A.W.A., Tam, V.W.Y. and Haddad, A. (2022). Life-cycle assessment methodology integrated with BIM as a decision-making tool at early stages of building design. *International Journal of Construction Management*, *22*(4), 541–55. https://doi.org/10.1080/15623599.2019.1637098.

Nehasilová, M., Lupíšek, A., Coufalová, P.L., Kupsa, T., Veselka, J., Vlasatá, B., Železná, J., Kunová, P. and Volf, M. (2022). Rapid Environmental Assessment of Buildings: Linking Environmental and Cost Estimating Databases. *Sustainability*, *14*(17), 10928. https://doi.org/10.3390/su141710928.

Newberry, P., Harper, P. and Norman, J. (2023). Carbon assessment of building shell options for eco self-build community housing through the integration of building energy modelling and life-cycle analysis tools. *Journal of Building Engineering*, *70*, 106356. https://doi.org/10.1016/j.jobe.2023.106356.

Page, M.J., McKenzie, J.E., Bossuyt, P.M., Boutron, I., Hoffmann, T.C., Mulrow, C.D., Shamseer, L., Tetzlaff, J.M., Akl, E.A., Brennan, S.E., Chou, R., Glanville, J., Grimshaw, J. M., Hróbjartsson, A., Lalu, M.M., Li, T., Loder, E.W., Mayo-Wilson, E., McDonald, S., …and Moher, D. (2021). The PRISMA 2020 statement: An updated guideline for reporting systematic reviews. *The BMJ*, *372*. https://doi.org/10.1136/BMJ.N71.

Parece, S., Resende, R. and Rato, V. (2024). A BIM-based Tool for Embodied Carbon Assessment using a Construction Classification System. *Developments in the Built Environment*, 100467. https://doi.org/10.1016/j.dibe.2024.100467.

Renovation Wave. (2024). Retrieved May 21, 2024. https://energy.ec.europa.eu/topics/energy-efficiency/energy-efficient-buildings/renovation-wave_en.

Röck, M., Saade, M.R.M., Balouktsi, M., Rasmussen, F.N., Birgisdottir, H., Frischknecht, R., Habert, G., Lützkendorf, T. and Passer, A. (2020). Embodied GHG emissions of buildings: The hidden challenge for effective climate change mitigation. *Applied Energy*, *258*, 114107. https://doi.org/10.1016/J.APENERGY.2019.114107.

Safari, K. and AzariJafari, H. (2021). Challenges and opportunities for integrating BIM and LCA: Methodological choices and framework development. *Sustainable Cities and Society*, *67*, 102728. https://doi.org/10.1016/J.SCS.2021.102728.

Santos, R., Costa, A.A., Silvestre, J.D., Vandenbergh, T. and Pyl, L. (2020). BIM-based life cycle assessment and life-cycle costing of an office building in Western Europe. *Building and Environment*, *169*, 106568. https://doi.org/10.1016/J.BUILDENV.2019.106568.

Schneider-Marin, P., Stocker, T., Abele, O., Margesin, M., Staudt, J., Abualdenien, J. and Lang, W. (2022). EarlyData knowledge base for material decisions in building design. *Advanced Engineering Informatics*, *54*, 101769. https://doi.org/10.1016/j.aei.2022.101769.

Serrano-Baena, M.M., Ruiz-Díaz, C., Boronat, P.G. and Mercader-Moyano, P. (2023). Optimising LCA in complex buildings with MLCAQ: A BIM-based methodology for automated multi-criteria materials selection. *Energy and Buildings*, *294*, 113219. https://doi.org/10.1016/j.enbuild.2023.113219.

Sharif, S.A. and Hammad, A. (2019). Simulation-Based Multi-Objective Optimization of institutional building renovation considering energy consumption, Life-cycle Cost and Life-cycle Assessment. *Journal of Building Engineering*, *21*, 429–45. https://doi.org/10.1016/J.JOBE.2018.11.006.

Shibata, N., Sierra, F. and Hagras, A. (2023). Integration of LCA and LCCA through BIM for optimized decision-making when switching from gas to electricity services in dwellings. *Energy and Buildings*, *288*, 113000. https://doi.org/10.1016/j.enbuild.2023.113000.

Sobhkhiz, S., Taghaddos, H., Rezvani, M. and Ramezanianpour, A.M. (2021). Utilization of semantic web technologies to improve BIM-LCA applications. *Automation in Construction*, *130*, 103842. https://doi.org/10.1016/J.AUTCON.2021.103842.

Soust-Verdaguer, B., Bernardino Galeana, I., Llatas, C., Montes, M.V., Hoxha, E. and Passer, A. (2022). How to conduct consistent environmental, economic, and social assessment during the building design process. A BIM-based Life Cycle Sustainability Assessment Method. *Journal of Building Engineering*, *45*, 103516. https://doi.org/10.1016/j.jobe.2021.103516.

Soust-Verdaguer, B., Gutiérrez, J.A. and Llatas, C. (2023). Development of a Plug-in to Support Sustainability Assessment in the Decision-making of a Building Envelope Refurbishment. *Buildings*, *13*(6), 1472. https://doi.org/10.3390/buildings13061472.

Soust-Verdaguer, B., Gutiérrez Moreno, J.A. and Llatas, C. (2023). Utilization of an Automatic Tool for Building Material Selection by Integrating Life Cycle Sustainability Assessment in the Early Design Stages in BIM. *Sustainability*, *15*(3), 2274. https://doi.org/10.3390/su15032274.

Soust-Verdaguer, B., Llatas, C. and García-Martínez, A. (2017). Critical review of BIM-based LCA method to buildings. *Energy and Buildings*, *136*, 110–120. https://doi.org/10.1016/J.ENBUILD.2016.12.009.

Stellacci, S., Domingos, L. and Resende, R. (2024). Integrated computational approaches for energy retrofit of historical buildings in extreme climate environments. *International Journal of Building*

Pathology and Adaptation, *42*(1), 114–32. https://doi.org/10.1108/IJBPA-03-2022-0044/FULL/XML.

Su, S., Li, S., Ju, J., Wang, Q. and Xu, Z. (2021). A building information modelling-based tool for estimating building demolition waste and evaluating its environmental impacts. *Waste Management*, *134*, 159–69. https://doi.org/10.1016/j.wasman.2021.07.025.

Taher, A.H. and Elbeltagi, E.E. (2023). Integrating building information modelling with value engineering to facilitate the selection of building design alternatives considering sustainability. *International Journal of Construction Management*, *23*(11), 1886–1901. https://doi.org/10.1080/15623599.2021.2021465.

Tushar, Q., Bhuiyan, M.A. and Zhang, G. (2022). Energy simulation and modelling for window system: A comparative study of life cycle assessment and life-cycle costing. *Journal of Cleaner Production*, *330*. https://doi.org/10.1016/j.jclepro.2021.129936.

Tushar, Q., Bhuiyan, M.A., Zhang, G. and Maqsood, T. (2021a). An integrated approach of BIM-enabled LCA and energy simulation: The optimised solution towards sustainable development. *Journal of Cleaner Production*, *289*, 125622. https://doi.org/10.1016/j.jclepro.2020.125622.

Tushar, Q., Bhuiyan, M.A., Zhang, G. and Maqsood, T. (2021b). An integrated approach of BIM-enabled LCA and energy simulation: The optimised solution towards sustainable development. *Journal of Cleaner Production*, *289*. https://doi.org/10.1016/j.jclepro.2020.125622.

Tushar, Q., Zhang, G., Bhuiyan, M.A., Navaratnam, S., Giustozzi, F. and Hou, L. (2022). Retrofit of Building Façade Using Precast Sandwich Panel: An Integrated Thermal and Environmental Assessment on BIM-based LCA. *Buildings*, *12*(12), 2098. https://doi.org/10.3390/buildings12122098.

Uddin, M.N., Wei, H.H., Chi, H.L., Ni, M. and Elumalai, P. (2021). Building information modelling (BIM) incorporated green building analysis: An application of local construction materials and sustainable practice in the built environment. *Journal of Building Pathology and Rehabilitation*, *6*(1), 13. https://doi.org/10.1007/s41024-021-00106-5.

United Nations Environment Programme. (2024). Global Status Report for Buildings and Construction: Beyond foundations: Mainstreaming sustainable solutions to cut emissions from the buildings sector. Retrieved November 22, 2024, from https://globalabc.org/resources/publications/global-status-report-buildings-and-construction-beyond-foundations

VOSviewer. (n.d.). *Visualizing scientific landscapes*. Retrieved May 27, 2024. https://www.vosviewer.com/

Wastiels, L. and Decuypere, R. (2019). Identification and comparison of LCA-BIM integration strategies. *IOP Conference Series: Earth and Environmental Science*, *323*(1), 012101. https://doi.org/10.1088/1755-1315/323/1/012101.

Xu, J., Teng, Y., Pan, W. and Zhang, Y. (2022). BIM-integrated LCA to automate embodied carbon assessment of prefabricated buildings. *Journal of Cleaner Production*, *374*, 133894. https://doi.org/10.1016/j.jclepro.2022.133894.

Zhang, Y., Jiang, X., Cui, C. and Skitmore, M. (2022). BIM-based approach for the integrated assessment of life cycle carbon emission intensity and life-cycle costs. *Building and Environment*, *226*, 109691. https://doi.org/10.1016/j.buildenv.2022.109691.

Zheng, B., Hussain, M., Yang, Y., Chan, A.P.C. and Chi, H.-L. (2023, July). Trade-offs between accuracy and efficiency in BIM-LCA integration. *Engineering, Construction, and Architectural Management*. https://doi.org/10.1108/ECAM-03-2023-0270.

Zhou, Y., Ma, M., Tam, V. W. and Le, K. N. (2023). Design variables affecting the environmental impacts of buildings: A critical review. *Journal of Cleaner Production*, *387*, 135921. https://doi.org/10.1016/J.JCLEPRO.2023.135921.

Zhou, Y., Tam, V.W. and Le, K.N. (2023a). Developing a multi-objective optimisation model for improving building's environmental performance over the whole design process. *Building and Environment*, *246*, 110996. https://doi.org/10.1016/J.BUILDENV.2023.110996.

Zhou, Y., Tam, V.W. and Le, K.N. (2023b). Sensitivity analysis of design variables in life-cycle environmental impacts of buildings. *Journal of Building Engineering*, *65*, 105749. https://doi.org/10.1016/J.JOBE.2022.105749.

Epilogue

The Future of Swarm Intelligence for Smart Cities and Beyond

As we draw this comprehensive exploration to a close, it is evident that the application of swarm intelligence (SI) and related digital innovations holds transformative potential for the cities of the future. This book has provided an extensive examination of various aspects where technology, social learning, citizen engagement, public policies, and innovative reporting intersect to shape urban environments that are sustainable, resilient, and inclusive.

Social Learning and Citizen Engagement

Chapter 1 and Chapter 2 have underscored the critical role of social learning and citizen engagement in the development of future cities. By harnessing collective intelligence through swarm intelligence, cities can become more adaptable and responsive to the needs of their inhabitants. The integration of technologies such as Extended Reality (XR), Artificial Intelligence (AI), and digital games demonstrates how these tools can facilitate transparent and inclusive decision-making processes. These innovations not only enhance citizen participation but also democratize governance, fostering a sense of community ownership and collaborative problem-solving.

The concept of social learning allows for the observation, retention, reproduction, and motivation of behaviors exhibited by others, thus enabling citizens to learn from each other and contribute meaningfully to urban development initiatives. Through the extensive literature review and analysis of case studies, it becomes apparent that the successful integration of these technologies can lead to the creation of urban environments that are not only more efficient but also more attuned to the needs and aspirations of their residents. This deepened understanding of social dynamics and technological applications showcases how future cities can leverage the collective wisdom of their populations to solve complex urban problems and create more harmonious living spaces.

Technological Facilitation and Public Policies

As discussed in Chapter 3, public policies play a vital role in guiding the digital transformation towards greener cities. The systematic review of policy dimensions and applied digital technologies reveals the importance of interpretive approaches in policy analysis. This understanding enables policymakers to craft strategies that align with the environmental and technological needs of urban areas. Moreover, the insights from Chapter 4 highlight the significance of innovative financial reporting mechanisms, like the *Popular Financial Report* (PFR), in enhancing transparency and accountability within smart city initiatives. These reports serve as pivotal tools for fostering trust between citizens and public institutions, thereby enhancing citizen engagement and promoting environmental sustainability.

The interaction of various stakeholders in policymaking processes underscores the collaborative effort required to address the complex challenges faced by modern urban environments. The integration of digital technologies in policymaking not only streamlines processes but also ensures that policies are more responsive to real-time data and changing circumstances, thereby making urban governance more dynamic and effective. The importance of technology as a facilitator in these processes cannot be overstated, as it bridges the gap between policy intentions and real-world applications, making policies more adaptable and impactful.

Public Safety and IoT Systems

Chapter 5 and Chapter 6 delve into the critical aspect of public safety and the role of IoT systems in enhancing urban living. The use of Urban Digital Twins, Blockchain, and real-time data collection through sensors and networks emphasizes the need for robust frameworks to mitigate the impacts of climate change. These technologies not only improve indoor environmental quality and public safety but also contribute to energy efficiency and comfort for city inhabitants. By providing a comprehensive framework that integrates early warning systems and monitoring Indoor Environmental Quality, stakeholders can access timely and accurate information for effective decision-making.

The case study of a neighborhood in Gaia, Portugal, exemplifies the practical implementation of these technologies, demonstrating how indoor sensors can measure environmental quality and enhance public safety at the local level. The broader adoption of such digital tools across future smart cities worldwide promises to pave the way for a more sustainable and resilient urban future despite environmental challenges. The detailed examination of various IoT architectures, including centralized, distributed, edge computing, and fog computing, highlights the diverse approaches available for optimizing urban systems and enhancing the quality of life for residents. These technologies, when implemented effectively, can significantly reduce the risks associated with climate change and other environmental challenges, ensuring that future cities remain safe and livable for their residents.

Energy Management and Community Engagement

The exploration of swarm intelligence in energy management, as presented in Chapters 7 and 8, showcase the potential for optimizing energy usage within buildings and communities. By leveraging the collective behavior of decentralized entities, cities can achieve significant energy savings and foster sustainable practices. The practical applications within European projects illustrate how swarm intelligence can support community energy balancing, demonstrating the feasibility and benefits of such approaches. This approach not only enhances the efficiency of energy consumption but also promotes the use of renewable energy sources, thereby contributing to the overall sustainability of urban environments.

The coordinated communication and collaboration among sensors, smart devices, and building systems enable real-time optimization of energy usage based on various factors such as occupancy, temperature, lighting, and weather conditions. This adaptive control strategy not only reduces energy consumption and costs but also ensures that urban infrastructures are better equipped to handle the demands of a growing population and changing environmental conditions. The potential for swarm intelligence to revolutionize energy management practices in urban settings highlights the importance of continued research and development in this field, as well as the need for collaborative efforts to implement these technologies on a larger scale.

Healthcare and Urban Resilience

Chapters 9 and 10 address the intersection of healthcare, sustainability, and urban resilience. The development of evidence-based tools for assessing sustainability and inclusion in healthcare infrastructures highlights the need for innovative solutions in this critical sector. The methodological evolution of these tools, developed within research projects, underscores their importance in providing comprehensive data collection and reliable analysis. By focusing on enhancing quality, sustainability, and user-centered experiences, these tools contribute significantly to the sustainable design of healthcare environments.

The comparative analysis of urban resilience in Milan and Rome provides valuable insights into how digital technologies and strategic planning can enhance the resilience of urban environments against climate change and other challenges. The implementation of efficient integrated strategic planning through digital information systems, technological innovations, and novel models of urban governance is crucial to meeting EU targets by 2030. The reflections on the influence of digital technologies in reshaping the future of these cities emphasize the importance of fostering a culture centered around sustainability and resilience. The adoption of such technologies and strategies not only enhances the capacity of cities to withstand and recover from adverse events but also ensures that urban environments remain vibrant and livable for future generations.

The integration of healthcare considerations into urban planning processes highlights the interconnected nature of urban systems and the need for a holistic approach to ensure the overall well-being of city residents. By aligning healthcare infrastructure with broader urban resilience strategies, cities can create environments that support physical, mental, and social health, thus promoting a higher quality of life for all inhabitants.

Finally, Chapter 11 explores the potential of Building Information Modeling (BIM) tools for assessing the embodied carbon and sustainability of buildings. This forward-looking approach underscores the necessity of integrating advanced digital tools to meet sustainability targets and improve the overall environmental performance of urban infrastructures. By providing a detailed analysis of the existing BIM tools and their applications, this chapter highlights the significant impact these technologies can have on the construction and operation of buildings.

The focus on embodied carbon assessment ensures that the environmental impact of buildings is considered throughout their lifecycle, from construction to demolition. This holistic approach to building design and operation is essential for creating sustainable urban environments that minimize their carbon footprint and promote long-term environmental stewardship. The adoption of BIM tools represents a significant step forward in the effort to create more sustainable and environmentally conscious urban spaces, and continued advancements in this area will be crucial to achieving global sustainability goals.

Embracing the Future

In conclusion, the chapters in this book collectively demonstrate the profound impact that swarm intelligence and related digital technologies can have on the cities of the future. By fostering social learning, enhancing citizen engagement, implementing robust public policies, and leveraging innovative technological solutions, urban environments can become more sustainable, resilient, and inclusive. The insights and case studies provided herein offer a roadmap for researchers, policymakers, and urban planners to harness the power of collective intelligence and digital innovation in shaping the cities of tomorrow.

As we continue to navigate the complexities of urban development in the digital age, it is imperative to embrace these transformative approaches to create urban spaces that prioritize the well-being and prosperity of all citizens. The integration of these diverse elements into urban planning and development not only addresses the immediate challenges faced by modern cities but also ensures that future urban environments are better equipped to handle the evolving needs of their populations.

By adopting a holistic and collaborative approach to urban governance, we can create cities that are not only more efficient and sustainable but also more inclusive and responsive to the needs of their inhabitants. The journey towards building the cities of the future is a continuous process that requires the active participation and engagement of all stakeholders, from policymakers and urban planners to citizens and technology developers.

By working together and leveraging the power of swarm intelligence and digital innovations, we can pave the way for a brighter, more sustainable, and resilient future for our urban environments. This collaborative effort is essential in ensuring that the cities of the future are not only technologically advanced but also socially equitable and environmentally sustainable. The vision of future cities presented in this book is one of harmony, where technology and human ingenuity come together to create urban spaces that enhance the quality of life for all residents.

As we move forward, it is crucial to remain committed to this vision and to continue exploring new ways to integrate swarm intelligence and digital technologies into the fabric of our cities, ensuring that they remain vibrant, resilient, and inclusive for generations to come.

The Broader Horizon: Integrating Swarm Intelligence Across Domains

The promise of swarm intelligence is not confined to urban environments alone. Its principles and applications have the potential to transform various other domains, thereby extending its benefits to wider societal contexts. The adaptability, scalability, and robustness of SI systems make them suitable for addressing complex problems across multiple fields.

Disaster Response and Management

One of the most promising applications of swarm intelligence is in disaster response and management. By employing swarms of drones or robots, emergency services can perform search and rescue operations more efficiently. These autonomous agents can cover vast areas, communicate with each other, and adapt to the changing conditions of disaster sites. Swarm intelligence enables these agents to work collectively without the need for central control, which is crucial in chaotic and unpredictable environments.

The real-time data collection and analysis capabilities of these swarms provide critical information to human responders, enhancing their situational awareness and decision-making processes. This technology can significantly reduce response times, increase the effectiveness of rescue operations, and ultimately save lives. As natural disasters become more frequent and severe due to climate change, the role of swarm intelligence in disaster management will become increasingly important.

Healthcare Innovations

In the healthcare sector, swarm intelligence can lead to groundbreaking innovations, particularly in the field of medical nanotechnology. Swarms of nanobots could be used for targeted drug delivery, precision surgery, and real-time health monitoring. These nanobots, operating collectively, can navigate the human body to perform complex tasks that are currently beyond the reach of conventional medical technologies.

For instance, in cancer treatment, nanobots can deliver drugs directly to tumor sites, minimizing the side effects on healthy tissues and improving the efficacy of the treatment. Additionally, swarms of nanobots can perform intricate surgical procedures with unprecedented precision, reducing recovery times and improving patient outcomes. The integration of swarm intelligence in healthcare not only promises to revolutionize medical treatments but also to enhance the overall quality of patient care.

Environmental Conservation

Swarm intelligence also holds significant potential for environmental conservation efforts. Autonomous swarms of drones or robots can be deployed for monitoring wildlife, managing natural resources, and restoring ecosystems. These agents can work together to collect data on environmental conditions, track the movements of endangered species, and implement conservation measures.

For example, swarms of drones can be used to plant trees in deforested areas, monitor air and water quality, and detect illegal activities such as poaching or logging. The ability of these swarms to operate autonomously and adapt to changing conditions makes them ideal for large-scale environmental monitoring and intervention. By leveraging swarm intelligence, conservationists can enhance the effectiveness of their efforts and ensure the long-term sustainability of natural ecosystems.

Agricultural Advancements

In agriculture, swarm intelligence can contribute to more efficient and sustainable farming practices. Swarms of autonomous robots can perform various tasks such as planting, weeding, and harvesting, optimizing the use of resources, and minimizing environmental impact. These robots can work collectively to monitor crop health, detect pests and diseases, and apply treatments precisely where needed.

The integration of swarm intelligence in precision farming can lead to higher crop yields, reduced use of pesticides and fertilizers, and more sustainable land management practices. Farmers can benefit from real-time data on soil conditions, weather patterns, and crop growth, enabling them to make informed decisions and improve their productivity. As the global population continues to grow, the adoption of swarm intelligence in agriculture will be essential for ensuring food security and sustainability.

Ethical and Social Considerations

While the potential benefits of swarm intelligence are vast, it is crucial to address the ethical and social implications of its deployment. The widespread use of autonomous systems raises important questions about privacy, security, and accountability. Ensuring that these technologies are developed and used responsibly is essential for maintaining public trust and preventing unintended consequences.

Privacy and Security

The use of swarm intelligence in urban environments, healthcare, and other domains involves the collection and processing of large amounts of data. Protecting the privacy of individuals and ensuring the security of this data is paramount. Robust data protection measures and transparent policies must be implemented to safeguard personal information and prevent misuse.

In addition, the security of autonomous systems themselves must be ensured. As these systems become more prevalent, they could become targets for cyber-attacks. Implementing strong cybersecurity measures and designing systems that are resilient to attacks is critical.

The exploration of swarm intelligence and its multifaceted applications across smart cities, healthcare, disaster management, environmental conservation, and agriculture illustrates a transformative potential that is both profound and far-reaching. This book has highlighted how the integration of advanced digital technologies and collective intelligence can foster urban environments that are not only more efficient and sustainable but also more inclusive and resilient. As we move forward, the ethical and social implications of deploying such technologies must be addressed to ensure that the benefits are maximized while minimizing potential risks. By embracing these innovative approaches and fostering a collaborative effort among all stakeholders, we can create a future where technology and human ingenuity work in harmony to enhance the quality of life for all citizens, paving the way for a more sustainable, resilient, and equitable world. The vision presented herein is a testament to the potential of swarm intelligence to revolutionize our urban spaces and beyond, promising a brighter future for generations to come.

Index

For Product Safety Concerns and Information please contact our EU representative GPSR@taylorandfrancis.com Taylor & Francis Verlag GmbH, Kaufingerstraße 24, 80331 München, Germany